Quirky Sides of Scientists

Quirky Sides of Scientists

True Tales of Ingenuity and Error
From Physics and Astronomy

David R. Topper

 Springer

David R. Topper
Department of History
University of Winnipeg
Winnipeg, Manitoba
Canada R3B 2E9
d.topper@uwinnipeg.ca

ISBN-13: 978-1-4419-2429-2   e-ISBN-13: 978-0-387-71019-8

Printed on acid-free paper.

9 8 7 6 5 4 3 2 1

springer.com

To Sylvia

You endured my obsession
with these "old dead friends" of mine
—while you took care of the living.

Acknowledgments

For their assistance and support at various stages of this project I kindly thank Wayne Choma, James Hanley, Scott Montgomery, and Dwight Vincent. For his congenial camaraderie spanning more than three decades, I am grateful to my much-missed late friend and colleague in the history of science, David Dyck. I am also beholden to all my students who asked (and continue to ask) challenging questions, forcing me to concede that I don't know many things—and compelling me to find them out. And, not the least, I acknowledge the University of Winnipeg, especially for granting me a research leave to complete this task.

Whatever singular insights may be found in this book, I will be pleased to accept the accolades for—though fully cognizant that all errors, too, are mine alone.

A Note on the Figures

I drew all 57 figures. Figures 3.1, 3.5, 4.1, 6.3, 6.4, 6.6, 8.4, and 11.4 are based on images from I.B. Cohen's *The Birth of a New Physics* (New York: W.W. Norton, 1960; revised ed.1985), a teaching text I have used for over three decades. Others are from articles I have published, which are cited at the end of the corresponding chapters. In my sketches of original historical images I have translated any Latin inscriptions into English. The consequential loss entailed in not reproducing the original diagrams and artwork is mitigated by the present-day fact that most images are available on the Internet.

Contents

Prelude

This is unabashedly an idiosyncratic look at science. Based heavily upon my research and publications—and hence personal interests—it expresses, perhaps, my quirky side.

Ever since switching fields in graduate school, from physics to studying its history, I have come to recognize—and especially appreciate—what a thorny matter it is for the apparently simple laws of science to immerge out of the shadows of history. This, in turn, made me realize how remarkable and unobvious the expositions in today's science textbooks are; what is eventually straightforward and transparent today was not so in the past. I did not appreciate the astonishing elegance of many science textbooks; that is, not until I studied the contrasting history of the subject. The subject of this book, however, is not textbook science; the textbook is the foil. A scientific idea, law, discovery, or experiment as explicated in a textbook is really a distillation of a multifaceted and often-intricate and convoluted historical narrative, with many missed starts, dead ends, mistakes, even sophistries and deceptions—a labyrinth of ingenuity and error that looks linear only after the fact.

The book thusly is directed at the reader who is curious about science but whose exposure has been primarily from science courses and their accompanying textbooks—thus devoid of real history. This book is also intended for those who take pleasure in carefully studying pictures, illustrations, diagrams, of which there are scores in this book, and who are willing to spend the time required to compare and comprehend figure and text, so as to follow the historical narrative and scientific argument. A significant segment of what purports to be writings on science directed beyond the classroom and academy is done so under the stricture that the text should read rather like a novel—as if, God forbid, one might need to stop and ponder a picture or think about an idea. I expect that anyone reading this book "like a novel" is wasting one's time.

The overall structure of the book is plain. Sandwiched between two chapters on the doggedness of Einstein are ten more in approximately chronological order, beginning with an ancient astronomical measurement followed by episodes from the Scientific Revolution (Copernicus through Newton), with accompanying background narratives from ancient science, telling tales revealing some quirky sides of scientists. The format is simple: every chapter constitutes more or less an essay unto itself; furthermore, each is divided into subsections, which are occasionally followed by boxed-in essays on specific peripheral topics. To preserve the integrity of the argument of each chapter while concomitantly avoiding too much duplication, I have cross-referenced relevant material from elsewhere in the book by section number, for example, "see section 2.3" means section 3 of Chapter 2. Accordingly, the book may be read in any order. So, begin browsing, perusing, reading—wherever you wish.

1
Tenacity and Stubbornness: Einstein on Theory and Experiment

One of the common images of Einstein is this: lost in thought, he scribbles esoteric equations on the back of an envelope and creates a new world. Out of the mathematics on the envelope emerges a theory that he is so convinced is true, no experiment is required for proof. There are many stories of Einstein's disdain for experiments, especially when the experiments contradicted his theories. But a deeper look into the matter reveals a more complex attitude toward experimentation. Especially interesting is a little-known tale of Einstein himself engaged in an experiment to test one of his own theories—and how he confronted his own data. One of the guiding factors in all this was Einstein's tenacity for sticking with a scientific problem, following wherever it leads, and stubbornly not straying from the quest.

1.1. Tenacity

When once asked how he discovered the laws of physics, Newton said it was straightforward—he just thought about the problem, constantly. Confronted with a problem, Newton became obsessed until he solved it. But most of his life was not spent cracking what today we would consider problems in science; indeed, his interest in science was infrequent. Newton mostly was occupied with theology, church history, and alchemy, although for him they were all of a piece.

Einstein, too, once he began, seemingly never stopped thinking about physics. But, in contrast to Newton, science was mostly what he pursued throughout his life. He had little commitment to his family, and except for his involvement with various social concerns (socialism, Zionism, the world-government movement), his life was spent zealously preoccupied with matters of physics. He tells us in his autobiography that at about the age of 16 he thought about a paradox concerning how light would appear if he traveled along with the light beam, and this imaginary ride led, about 10 years later, to what we now call the special theory of relativity. The resolution of the paradox entailed two principles: the relativity of motion and the invariant speed of light. In that year, 1905, he not only published two papers on relativity (in the second deducing $E = mc^2$) but also laid the foundations of the quantum theory of light (from the photoelectric effect and other light-related phenomena) and provided further evidence for the existence of atoms, whose real existence was neither fully confirmed nor held at the time. With these accomplishments, he could have rested on his laurels.

But not Einstein. Special relativity was confined to the case of motion where an observer moves at a constant speed in a straight line (later labeled an "inertial system"). But what about the more general case of an observer changing speed, or accelerating? How may the relativity principles apply to this "noninertial system"? While writing a summary article on relativity in 1907, a thought occurred to him, which offered an answer. The thought was this: the physical experience of a person in free-fall in a gravitational field is the same as if gravity were turned off. Although the person is falling, and hence accelerating, her experience is identical to someone floating in outer space, away from any gravity. And, concomitantly, if a person is in a local environment without gravity (image a spaceship far from any other masses), but it is accelerating (thus being a noninertial system), he experiences a force identical to being in a gravitational field (Fig. 1.1).

Our experience in an elevator readily reproduces this: as the elevator goes up, we feel heavier, as if gravity were increased; the opposite occurs as the elevator goes down—we feel lighter. An extreme case would be an elevator in free-fall, where we would be weightless, as if floating in free space. Similarly, someone in an elevator in free space that is accelerating "up" with respect to the top would experience the equivalence of a gravitational force. Interestingly, the modern elevator first appeared in the mid-19th century; hence the experience perhaps was familiar to Einstein from his childhood.

From this famous thought experiment Einstein concluded that gravitational force might ultimately be a form of inertial force, since the person in the accelerating

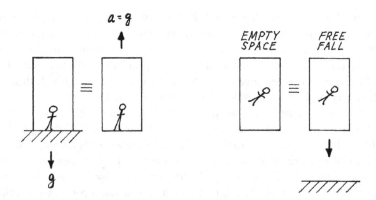

FIGURE 1.1. The equivalence of gravitational and inertial forces. Left: Experiencing a gravitational field is equivalent to being in a noninertial frame of reference that is accelerating (a) at the same rate (distance/time-squared) as gravity (g). Right: Experiencing weightlessness in empty space is equivalent to falling freely, hence accelerating, in a gravitational field.

elevator is experiencing the equivalence of gravity because of her inertia—that is, her mass is resisting the change of speed. Another way of saying this is that inertial mass is equivalent to gravitational mass. That there is proportionality between inertial mass and gravitational mass was inferred by Galileo and Newton from physical experiments, but Einstein's thought experiment revealed something more fundamental. It permitted him to generalize the relativity principle to all cases of motion (inertial and noninertial systems), and hence elevate the equivalence notion to an *a priori* "principle." He later called this "the happiest thought of my life."

He pondered and developed the consequences of this for nearly the next decade. The principle of equivalence (between inertia and gravity) was the cornerstone of what he called the general theory of relativity, which he completed and published late in 1915, producing a now-classic review article in 1916. The theory posited that gravity is not a force *in* space (as Newton said) but a property *of* space. To Newton gravity was an example of what the ancients called action-at-a-distance—namely, the ability of one mass to transmit information instantaneously across space to another mass. Einstein once called action-at-a-distance "spooky." Instead, he proposed that space itself is the cause of gravity; the ability of space to wrap, warp, or bend around matter is what gives rise to masses in space moving toward each other; gravitational attraction, therefore, is nothing more than the distortion of space. Many theorists have deemed Einstein's general relativity as the most beautiful theory ever conceived. Pictorially it is relatively simple to conceive in a two-dimensional world: for the two-dimensional "person" in Figure 1.2, the distortion of space around the large mass is interpreted as a

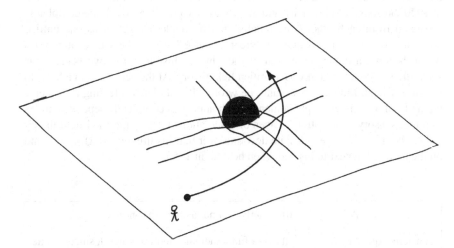

FIGURE 1.2. Warped space: two-dimensional analogue. Matter distorts the immediate space around it. For the 2D "person," the behavior of the moving smaller mass as it curves around the larger mass is perceived as being due to a force acting between the masses, and thus a gravitation attraction is postulated. Einstein, accordingly, reduced gravity to the curvature of space.

force, since the thrown object seems to be attracted to the large mass and hence the "person" posits a force between them.

Likewise, for us, gravity, therefore, is due to a distortion of our three-dimensional space. In contrast to the conception, the mathematical execution was formidable, requiring the mastering of tensor calculus in order to describe motion in a distorted (technically non-euclidean; see section 12.2) space. That Einstein thus explained gravity without spooky forces was perhaps why he called the equivalence principle the happiest thought of his life.

After years of calculating tensor equations and achieving the general theory, Einstein may well have rested for a while. But not he: the very next year (1917) we find him publishing what eventually will be another landmark paper—this one on a model of the universe as inferred by general relativity. This cosmological model became the foundation of modern cosmology (see section 12.2).

Reaching this plateau, would Einstein now take a break? Indeed, what more was there to do? Well, tucked deep within the 1916 review article on general relativity was the notion of merging gravity with electricity under one theory. Over 30 years later, in his autobiography, Einstein recalled that this thought arose out of his derivation of the field equation of general relativity. "Not for a moment," he wrote, "did I doubt that this formulation [of general relativity in 1915–16] was merely a makeshift [one]. . . . For it was essentially *no more* than a theory of the gravitational field, which was isolated somewhat artificially from a total field of as yet unknown structure." There was more to the physical universe than gravitational force alone, since there were also forces of electricity and magnetism. This led to his quest for the unification of gravitation and electromagnetism—to find, as he called it, this total field of "unknown structure." In a lecture delivered at the University of Leiden in 1920, he spoke of electricity and gravity as "two realities which are completely separated from each other conceptually" but if coupled "together as one unified conformation" would constitute a "great advance," since "the whole of physics would become a complete system of thought." A fusion of the two became his scientific mission; the goal was a unified field theory. At the age of 22, in a letter to his close friend, Marcel Grossmann, he wrote: "It is a glorious feeling to perceive the unity of a complex of phenomena which appear as completely separate entitles to direct sensory observation." Pages of calculations toward a unified field theory (which he never achieved, despite devoting over half his life toward it) were found on the night table next to his bed when he died in 1955.

A Unity in Mind, Only: Linnaeus and Taxonomy

Einstein's quest remains just that—a fundamental scientific pursuit, since no one has yet found the unity between gravity and electricity. Perhaps there is none, although few physicists would agree (an exception is Freeman Dyson, professor emeritus of physics at the Institute for Advanced Study). The story in Chapter 9 tells of Newton's misguided quest for a unity among color, light, and sound.

Here is another example, this from the history of taxonomy. A landmark publication on the subject is the *Systema Naturae* of Carl Linnaeus. From the first edition in 1735 and proceeding through 12 editions (the last in 1766–8) Linnaeus's masterpiece laid the groundwork for the modern system of classifying all living things. His categories were, starting with the largest: kingdom, class, order, genus, and species. Phylum and family were added by Georges Cuvier in the early 19th century. Within this structure, each category branches out into further subcategories, never to be linked again. To Linnaeus this order was the rational framework in the mind of God, an arrangement based on the similarities and differences among the various species created by the Deity. Darwin subsequently transformed this logical ordering into a chrono-logical sequence, viewing the structure as a tree; hence, the metaphor entailed the branching as being literal.

In the latter editions of the *Systema,* Linnaeus extended this taxonomic structure to include rocks and human diseases. In antiquity Aristotle had divided nature into three kingdoms, animal, vegetable (plants), and mineral (rocks). Linnaeus, likewise, conceiving these three as part of the whole of nature, and, drawing on the belief that God would only create the diversity of nature within some *one* overall array, Linnaeus also classified rocks and even diseases as similar branching categories (which, of course, is erroneous since this "order" only applies to organic things because the basis of the sequence is evolution). Nevertheless, within the 18th century framework of a static picture of nature, the holistic framework of Linnaeus's fusion of rocks with animals and plants was reasonable.

1.2. Stubbornness

When convinced of a theory, Einstein could be very stubborn—even in light of seemingly empirical falsification. In March 1914, when in the midst of calculations involving his theory of general relativity, he wrote to another good friend, Michele Besso, on the possibility of testing his theory by measuring the bending of the light from stars during a solar eclipse: "Now I am completely satisfied and no longer doubt the correctness of the whole system, regardless of whether the observation of the solar eclipse will succeed or not. The logic of the thing is too evident." Usually he was right about these hunches—we are speaking of Einstein!—and so historians and others are prone to tell and retell those typically "Einsteinian" stories and anecdotes in which he disdained the role of experiment.

One of the most repeated anecdotes is this. The solar eclipse test of general relativity was finally performed under the direction of the Royal Society of London in 1919, and with a positive result—namely, the light from the stars near the sun was bent an amount within the experimental error of Einstein's prediction. A student of Einstein's at the time reports that she queried him as to the possibility of theory not being confirmed by the solar eclipse. To which, she reports, he responded: "Then I would have been sorry for the dear Lord. The theory is correct." Perhaps because of the invocation of the Creator, writers on Einstein have been drawn to this quote—indeed, here I am telling it, again. Moreover, it is seemingly reinforced by other statements, such as that to Besso (quoted above). But I must also report that this "dear Lord" story most probably is a myth. In a meticulously researched study of the story, Klaus Hentschel has convinced me the event never happened, and that it was a reconstruction by the student much later than the event (in the 1950s) to support her view that Einstein shared her philosophical (neo-Kantian) viewpoint (which he did not).

In his important article, Hentschel also makes a strong case for a modification of the widespread notion of antiempiricism in Einstein's ideology. There were numerous occasions where he clearly asserted a crucial role for experiments to make or break a theory; in fact, he recognized that they could be and often were the ultimate arbiters of scientific certainty. So, despite Einstein's remark to Besso (above), we find him as early as 1911, when he first derived the prediction that gravity bends light, exerting his influence to find someone willing to perform the solar eclipse test of the theory.

Hentschel especially emphasizes the role of a cluster of experiments (rather than one isolated case) as important to Einstein. Listen, for example, to this sentence from the introduction to his landmark paper of 1905 on his "quantum" (later photon) theory of light, a paper he referred to at the time as "very revolutionary":

Indeed, it seems to me that the observations of "blackbody radiation," photoluminescence, production of cathode rays by ultraviolet light, and other related phenomena associated with the emission or transformation of light appear more readily understood if one assumes that the energy of light is discontinuously distributed in space.

Note the list of phenomena (the details are not of interest here) supported by experiment, leading Einstein to his hypothesis. Other such examples abound in his writings.

On the other hand, one experiment alone may not carry much weight. Before an experiment may be accepted as constituting the final test of a theory, it too must be scrutinized for possible flaws in design and execution. A case in point involved, indeed, the first experimental test of the special theory of relativity. The one prediction of the theory that was possible to test around 1905 was an increase in the mass of matter moving near the speed of light. At the time experiments on "beta rays," which were identified as very fast electrons, were being performed in Germany by Walter Kaufmann; in 1906 he published the results of several years of experiments, and they showed that the measured mass of the fast electrons did increase, although not by the amount predicted by Einstein. Kaufmann interpreted his experiments as thus disproving Einstein's theory.

About this time Einstein was writing the review article on relativity, mentioned above. It was commissioned by the experimentalist and Nobel Prize winner Johannes Stark for a journal he edited. (Stark would later become a major nemesis, being a vitriolic anti-Semitic Nazi, accusing Einstein of corrupting pure-Aryan physics with his "Jewish physics.") In his manuscript for the journal, Einstein stressed, as he wrote to Stark, the importance of "intuitiveness and simplicity of the mathematical developments" of the theory, thus putting forth what later became his familiar emphasis upon the role of aesthetics in evaluating a theory. So when the article was published in 1907, he commented upon Kaufmann's experiments this way: "Whether there is an unsuspected systematic error or whether the foundations of relativity theory do not correspond with the facts one will be able to decide with certainty only if a *great variety* [*mannigfaltigeres,* my italics] of observational material is at hand." In short, experiments (a cluster of them) can overthrow a theory; but one alone is not sufficient. His guess of systematic error turned out to be correct, but not until 1916, when experiments by others confirmed Einstein's theory.

Seeing and Knowing

Einstein is notorious for seizing on the internal consistency of a theory—or, in a sense, its beauty—as a decisive criterion for assessing its validity. In a famous story, Werner Heisenberg reports on a meeting with Einstein in 1926, shortly after Heisenberg had formulated his theory of quantum physics, in which he purposely eliminated a physical model of the atom with electrons in orbit, since they are unobservable entities and hence only hypothetical. In so doing, he believed he was following the principle set forth by Einstein in his theory of relativity of only positing observable magnitudes. But Heisenberg was jolted by Einstein's reply; Einstein said that it was wrong to avoid unobservable magnitudes, since it is "the theory which decides what we can observe." From a psycho-perceptual viewpoint Einstein was right: what we know can influence what we see.

Here is a stark example involving geology from Charles Darwin. In his autobiography he recalls a trip to Wales with the geologist Adam Sedgwick during his student years at Cambridge. The year is 1831, about a decade before Louis Agassiz put forward his theory of the movement of glaciers and the idea of ice ages. Darwin writes: "This tour was of decided use in teaching me a little how to make out the geology of a country. . . . On this tour I had a striking instance how easy it is to overlook phenomena, however conspicuous, before they have been observed by anyone. We spent many hours in Cwm Idwal, examining all the rocks with extreme care, as Sedgwick was anxious to find fossils in them; but neither of us saw a trace of the wonderful glacial phenomena all around us; we did not notice the plainly scored rocks, the perched boulders, the lateral and terminal moraines. Yet these phenomena are so conspicuous [in retrospect] that . . . a house burnt down by fire did not tell its story more plainly than did this valley." Darwin put this succinctly once in a letter saying that "without the making of theories I am convinced there would be no observations."

Speaking of missing something in front of one's eyes, here's another example also involving Darwin. His and Alfred Russel Wallace's theories of evolution were presented at the meeting of the Linnean Society on July 1, 1858. At the end of the year the president's report contains this sentence: "The year . . .has not, indeed, been marked by any of those striking discoveries which at once revolutionize, so to speak, the department of science on which they bear."

A final example, this one from astronomy. The realization of the role of knowledge in perception, particularly as mediated by a telescope, was put by way of an analogy with music by William Herschel. He writes: "Seeing is in some respects an art which must be learnt. To make a person see with such a power [through my telescope] is nearly the same as if I were asked to make him play one of Handel's fugues upon the organ. Many a night I have been practicing to see, and it would be strange if one did not acquire a certain dexterity by such constant practice." The mention of Handel is not surprising, since Herschel, like George Frideric Handel, was initially also a German musician working in England; Handel made his mark early in the 18th century, and Herschel arrived in midcentury, but with his discovery of the planet Uranus he received a regular salary from the king and was able to devote his entire time to his hobby, astronomy.

1.3. Einstein's Experiment

There is another story of Einstein stubbornly ignoring empirical data that was contrary to his theory. This one is a little more complex and perhaps more interesting than the others, since here Einstein himself was involved in performing the experiment. The circumstances around this experiment, known as the Einstein–de Haas effect, took place in Berlin, early in 1915. It was a significant period in his life, for several reasons. He had moved to a prestigious position in Berlin in the spring of 1914, returning to the country he had departed, supposedly forever, at the age of 15. He moved with his wife and two sons, only to have his wife soon return to Zurich with their sons as she commenced divorce proceedings. In August, World War I began, with Einstein, although living in Germany, essentially rooting for the other side. Most important to his scientific life, in 1915 he was (as we now know) in the final stages of his calculations of the general theory of relativity, which he completed near the end of the year. Thus it is rather surprising finding him involved in, of all things, performing an experiment on electromagnetism. Maybe it was a distraction from the tedium of calculating seemingly endless tensor calculus equations.

There also was a personal factor. Einstein was a close friend of the Dutch physicist H.A. Lorentz, whose son-in-law, W.J. de Haas, had a temporary appointment (1913–1915) at a laboratory in Berlin; apparently, as requested by Lorentz, the joint experiment was undertaken partly as a sort of make-work project. De Haas had worked on magnetism in Leiden and, as will be seen, the experiment had relevance to atomic matters of interest to Einstein. And so he took de Haas under his wing, with the result being the Einstein–de Haas effect. (This has nothing to do with the Haas effect in physiology, which is about how we hear the spacing of sounds.)

The essence of the effect can be easily understood by a simple phenomenon. Consider a garden hose wound up on a reel and connected to a water source. When turning on the tap, a sort of "kick" (really a torque) is produced as the water winds its way through the hose; as a result, the hose rotates on the reel in the opposite direction of the flowing water. It is a case of Newton's action/reaction law. Einstein proposed that the same thing should occur when an electric current flows through a coil of wire. He devised this experiment to test it: An iron rod is suspended on a thin wire, with the rod hanging vertically through an electric coil; when an electric current flows through the coil, the induced magnetism in the iron should make the suspended rod rotate—like the water in the hose reel. Einstein predicted a value for the rotation from the laws of mechanics and electromagnetism. The prediction was $g = 1$, called the gyromagnetic ratio.

For the interested reader, here is a simple version of the calculation. Consider an electron of charge e and mass m, rotating in circle of radius r with velocity v and period T. The "current" is, by definition, e/T. Since $v = 2\pi r/T$, then the current is $ev/2\pi r$. The magnetic moment (M) is the current times the area orbit; namely, $ev/2\pi r \times \pi r^2 = evr/2$. The angular momentum ($A$) is mvr. By definition, the gyromagnetic ratio (g) is the ratio (M/A); that is, $e/2m$, which is set to unity. Therefore, the prediction was $g = 1$ for an orbiting electron.

The French scientist André-Marie Ampère, who made several important contributions to electromagnetic theory, had proposed a similar effect in the early 19th century, since he conceived of magnetism as due to the circular motion of electricity. Indeed, several failed attempts at measuring the effect were made throughout the 19th century.

Although reasonably simple in principle, in execution the experiment was quite thorny. Einstein and de Haas began working on it late in 1914, for we find Einstein writing to his friend Paul Ehrenfest in early December, "I am just in the process of starting an interesting experimental investigation with de Haas." Einstein wrote to his friend Besso in February: "The experiment will be coming to an end soon. With it the existence of zero point energy has also been proven in a single instance. A wonderful experiment; what a pity that you can't see it. And how treacherous (*heimtükkisch*) nature is, when you want to deal with it experimentally! Experimenting is becoming a passion for me even in my old age." ("Old age"? Einstein was not quite 36!) Zero-point energy was an idea of Max Planck's that Einstein was particularly interested in. Based on quantum theory, Planck proposed that an atom at absolute zero still possessed energy. Einstein believed that an orbiting electron was a model for this zero point energy. Moreover, this had relevance to a key issue in Niels Bohr's atomic model of 1913, namely that an electron moves in a fixed orbit around the nucleus, even though classical physics predicts that an orbiting (and hence accelerating) electron should radiate energy and in time spiral into the nucleus. These were fundamental problems at the time, so perhaps the experiment was more than a diversion from tedious tensor calculus calculations.

Einstein and de Haas finished the experiment early in 1915, before de Haas left Berlin in April. They performed two sets of experiments, obtaining $g = 1.45$ and $g = 1.02$; since the latter was close to that predicted by theory, the first they discarded. The second value was published in 1915. Bohr was very pleased; he saw the result as "direct support" for his model.

Subsequent experiments by others over the next few years, however, resulted in values for g around 2. Einstein insisted $g = 1$, as predicted by theory. But he was wrong. The correct value for g is, in fact, around 2, and the reason was not known until the early 1920s, with the discovery of what became known as electron "spin." It is the angular momentum of the electrons themselves, not their orbital motion, that is the cause of the gyromagnetic ratio, and this spin theory deduces a value for g around 2 (that is, $M/A = e/m$). Ironically then, if Einstein and de Haas had not thrown away their first experimental value, and perhaps performed further experiments, it is possible they may have reconsidered their theory and even predicted something like electron spin. But Einstein's stubbornness prevailed.

Einstein's Curious Sentence

Einstein's thought experiment about riding a beam of light is known from only one source, his autobiography, drafted in 1946. Putting pen to paper in his late 60s, he recalled this event from about the age of 16. The passage has become famous, and efforts to interpret it abound, but one point has been somewhat overlooked—the peculiarity of the last sentence. Let me begin by quoting the entire passage, minus that sentence.

If I pursue a beam of light with the velocity c (velocity of light in a vacuum), I should observe such a beam of light as an electromagnetic field at rest though spatially oscillating. There seems to be no such thing, however, neither on the basis of experience nor according to Maxwell's equations. From the very beginning it appeared to me intuitively clear that, judged from the standpoint of such an observer, everything would have to happen according to the same laws as for an observer who, relative to the earth, was at rest.

So far, so good. We see Einstein setting forth a contradiction between riding a beam of light at (obviously) the *known* speed of light and the principle of relativity, the latter implying that one's absolute motion cannot be experimentally measured. Einstein will resolve this paradox about 10 years later in his first paper on what became his special theory of relativity (1905) by postulating both the principle of relativity (for inertial systems) and the constant speed of light (in a vacuum, and independent of the motion of the source).

These two postulates ultimately formed the essential starting points for all of relativity, and hence their importance loomed large. Now, what does Einstein say in conclusion to the above argument? Here it is:

For how should the first observer know, or be able to determine, that he is in a state of fast uniform motion?

What? The observer is *not* supposed to know or determine his absolute motion, since that would contradict the principle of relativity. Whether at rest or moving in a straight line at constant speed, the observer should experience the same phenomenon. Moreover, why is Einstein ending this argument with a question? Rather, Einstein ought to be saying something like this: "For otherwise the observer will determine his absolute motion and this contradicts the principle of relativity." How could Einstein be so confused about such a historically and conceptually important linchpin of his theory?

When I noticed this conundrum, the obvious avenue I first explored was the translation of the passage from the original. Here is the German:

Denn wie sollte der erste Beobachter wissen bezw. konstatieren können, dass er sich im Zustand rascher gleichförminger Bewegung befindet?

The translation seems proper.

Being satisfied that the translation was accurate, the thought occurred to me that a mistake was made in the transcription from Einstein's original handwritten manuscript. With the help of Gerald Holton of Harvard University I obtained from the ongoing Einstein Papers project a copy of the requisite page in Einstein's hand. But it proved disappointing. Except for a minor correction by Einstein—apparently when writing the next to last word, he started writing *Beobachter* (observer) again but then wrote over what he had written and changed it to *Bewegung* (motion). So the transcript was correct.

I am thus left with a puzzle: how could Einstein misconstrue this important argument that was fundamental to the genesis of his theory of relativity? Holton agreed and pointed me toward an essay by the late Banesh Hoffmann, who likewise had noticed Einstein's curious sentence; Hoffmann called it "garbled."

I find it difficult to believe that Einstein would make such a crucial error in writing on this matter, but there seems to be no other explanation. I guess some questions just do not have reasonable answers.

Notes and References

Einstein's papers of 1905 are in vol. 2 of *The Collected Papers of Albert Einstein* (Princeton, NJ: Princeton University Press, 1987); the relativity review article is in vol. 2, Doc. 47. For the letters of Einstein: to Grossmann (April 1901), vol. 1, Doc. 100; to

Stark (November 1907), vol. 5, Doc. 63; to Besso (March 1914), vol. 5, Doc. 514; to Ehrenfest (December 1914), vol. 8, Doc. 39; to Besso (February 1915), vol. 8, Doc. 56.

In a letter to Max Born in 1947, Einstein called action-at-a-distance "spooky" (*spukhafte*): *The Born–Einstein Letters,* ed. Max Born and trans. Irene Born (London: MacMillan, 1971; new edition, 2005), p. 158, (new edition) p. 155.

The 1920 Leiden lecture is reprinted in Albert Einstein, *Sidelights on Relativity* (New York: Dover, 1983), pp. 1–24, quotation from pp. 22–23.

Klaus Hentschel, "Einstein's Attitude Towards Experiments: Testing Relativity Theory, 1907–1927," *Studies in History and Philosophy of Science,* 23, No. 4 (1992), pp. 593–624.

For the Einstein–de Haas effect see Peter Galison, *How Experiments End* (Chicago and London: University of Chicago Press, 1987), Chapter 2; and Abraham Pais, *"Subtle is the Lord": The Science and the Life of Albert Einstein* (New York: Oxford University Press, 1982), pp. 245–249.

Heisenberg's story (from his *Physics and Beyond* [1971]) is quoted in Gerald Holton, *The Scientific Imagination: Case Studies* (Cambridge: Cambridge University Press, 1978), pp. 216–217. I have used Charles Darwin, *The Autobiography of Charles Darwin,* ed. Nora Barlow (New York: W.W. Norton, 1969), p. 70. The quotation of the president of the Linnean Society is from Jonathan Howard, *Darwin* (Oxford: Oxford University Press, 1982), p. 6. The quotation from the letter (Darwin to Charles Lyell in 1860) is from David R. Stoddart, "Darwin and the Seeing Eye: Iconography and Meaning in the Beagle Years," *Earth Sciences History,* 14, No. 1 (1995), 3–22, on p. 5. The Herschel quotation is from, Richard Panek, *Seeing and Believing: How the Telescope Opened Our Eyes and Minds to the Heavens* (New York: Viking Press, 1998), p. 114.

Einstein's autobiography was written in 1946 and first published in 1949. I have used the corrected version, *Albert Einstein: Autobiographical Notes,* trans. Paul Arthur Schilipp (LaSalle and Chicago: Open Court Publishing, 1979), beam of light thought experiment on pp. 49–51. Banesh Hoffmann, "Some Einstein Anomalies," in Gerald Holton and Yehuda Elkana, eds., *Albert Einstein: Historical and Cultural Perspectives* (Princeton, NJ: Princeton University Press, 1982), 91–105; Hoffmann's remark on Einstein's "garbled" sentence is in footnote 3 on p. 105.

2
Convergence or Coincidence: Ancient Measurements of the Sun and Moon—How Far?

If two experiments measure the same phenomena from two different viewpoints and both get the same result (within, of course, experimental error) it stands to reason they are converging toward the correct answer. Why would anyone think otherwise? This chapter discusses the modern convergence of several measurements of the speed of light and compares this case with the ancient converging measurements of the relative distances of the moon and sun from Earth. The former was a real convergence. The latter was a coincidence. Could scientists at either time have known which is which?

2.1. The Speed of Light

Enter a dark room. Light a candle and the room seems instantaneously bright, dispelling the darkness. But is it instantaneous? One thing we can conclude from observation: the light is not evenly distributed over the room because it is brighter or more concentrated closer to the candle. This means that the light's intensity decreases with recession (the distance, D) from the source; that there is a mathematical law for this was first worked out by the German mathematician-astronomer Johannes Kepler in the 17th century. He showed that the intensity (I) of light obeys an inverse-square law (that is, $I \propto 1/D^2$). This seemed to imply that light is not instantaneous—for, if it were, the room would be uniformly lit. But can it be proven that the speed of light is finite?

The question about the possible instantaneous speed of light was debated over the ages, at least since the ancient Greeks. Little changed on this matter in the Middle Ages and Renaissance. Kepler's Italian contemporary, Galileo, tried to test it. Placing himself and an assistant on two distant hilltops at night with lanterns, he thought that by opening one's lantern when seeing the other's flash would result in a visible delay if the speed were finite. But the experiment was inconclusive: there was no time lag, which meant either that light moves instantaneously or the speed is too fast to measure by this experiment. In his last book, *Two New Sciences* (1638), Galileo wrote: "If not instantaneous, light is very swift."

Galileo, nevertheless, with his discovery of the moons of Jupiter in January 1610, bequeathed another method of measuring the speed of light. A study of the motions of the moons revealed that when Jupiter's moons passed behind the planet, the measured intervals of time were different at different times of the year, which corresponded to Jupiter being at different distances from Earth. Over three decades after Galileo died, the Danish astronomer Olaf Rømer concluded that this difference seemed to point to a time lag in the light from the moons reaching Earth—the sort of thing Galileo was looking for if the speed is finite. The Dutch scientist Christiaan Huygens used Rømer's data to calculate a value for the speed of light. The details of the experiment are not of concern here, only the result: the value was 124,000 miles/sec, which is extremely fast. This was the first empirical evidence that the speed of light is finite. The only question was the accuracy of this value.

In the early 18th century the Englishman James Bradley measured the speed of light from a different point of view, based on his discovery of what he called the aberration of light. He realized that in looking at stars through a telescope he had to adjust the angle of the sighting to take into account the motion of Earth; this was analogous to tilting an umbrella when walking through rain in order to keep dry. Again the details of the experiment are not of interest, only the value for the speed of light, which was186,233 miles/sec. Since both his and Huygens's values were within the same broad range, these two measurements were assumed to be "correct" in that the speed of light was indeed finite and that the task at hand was to measure it to closer and closer accuracy. This conviction was based on the fact that both numbers were arrived at by entirely different means, so it was reasonable to assume that these scientists were in fact measuring the same thing in different ways—that is, it was case of the convergence of data. Both measurements had been astronomical, and it is not surprising that the first measurements would be so, given the extreme speed of light.

In the next century the certainty in convergence was reinforced by some "tabletop" experiments. In mid–19th century the Frenchman H.L. Fizeau performed a laboratory experiment on light using a rotating cogwheel and got 194,000 miles/sec. Later the American Albert Michelson used a revolving mirror and, performing the experiment on several occasions, obtained an average value of 186,281 miles/sec.

Today the value is 186,290 miles/sec (299,790 km/sec). There is no doubt that in these cases of measuring the speed of light by different means, there *was* a convergence around and toward the correct value. As well, the speed of light is the fastest propagation of anything that we know of. In Galileo's experiment, the hills were only about a mile apart: no wonder his experiment was inconclusive!

This story of the measurement of the speed of light is by way of prefacing the account of a similar convergence of measurement made in ancient astronomy. This was a measurement of the relative distances of the sun and moon from Earth. To tell this tale I need to sketch some background information about everyday naked-eye astronomy and the ancient model of the universe.

2.2. Ancient Astronomy: Ptolemy

If science is, as purported to be, firmly rooted in careful observation, a systematic accumulation of empirical data, and the ability to predict further observations, then the ancient astronomers got it right. We now know that even in prehistoric times we humans carefully studied the motions of the heavens, especially the sun and moon, and occasionally a planet such as Venus (see section 4.1). Archaeological sites throughout the world attest to this seemingly fundamental human quest: Stonehenge in England, Chichén Itzá and other Mayan temples in the Yucatán peninsula, Medicine Wheels of the Aboriginal people of North America, and many more—their alignments to celestial events mark significant times of the year, such as solstices and equinoxes. In the Western tradition, beginning with the Babylonians and proceeding through the Greeks and Romans, we find a systematic study of the motions of the heavens, with further emphasis on the planets Venus, Mars, Jupiter, Saturn, and Mercury, which are visible to the naked eye.

A careful and systemic study of these motions, in all their details, permitted ancient astronomers to predict forthcoming celestial events. As well, the same naked eye observations were coupled to a commonsense view of the cosmos. If we accept the aphorism, "If something looks like a duck, walks like a duck, and quacks like a duck, then it's obviously a duck," then the cosmos is as the ancients pictured it. Their perception was their conception: the stars are fixed to a celestial sphere that rotates daily east to west around the central Earth. (Note how we use a hemisphere in a planetarium to reproduce an illusion of the sky; lights projected onto the interior of the dome look like stars because the night sky looks like that.) The moon, sun, and all the planets are carried by this motion and therefore rotate around Earth, daily rising in the east and setting in the west. But they also have a slower counter motion—(west to east) their intrinsic motion—and hence these bodies appear to back up, west to east, "through" the stars on regular cycles called sidereal periods (from the Latin, *sidereus*, pertaining to the stars). The fastest is the moon's sidereal period of 27 days. The sun's sidereal period (let's call it the solar year) is 365 days (in late antiquity this was fine-tuned to $365^{1}/_{4}$). The moon also goes through a cycle of phases every $29^{1}/_{2}$ days (crescent, full, etc.), the only body visibly doing so. The latter two cycles have been and are conveniently used by many cultures to divide the year into what we call a calendar.

We know, of course, that the changing phases of the moon are due to the relative positions of the moon and the sun, since the moon shines mainly by light reflected from the sun. But this was not obvious in antiquity; it is an inference drawn from observation. Just as the sun produces its own light (although this was not ubiquitously believed, either), so the moon may likewise generate its own light. It was alternately thought that the moon was two-toned (half black and half white), the phases thus being caused by a monthly rotation. (Note that this erroneous assumption entails a correct one; namely, that the moon is a sphere, which likewise is not obvious by mere observation.) Eventually a careful study of the moon's phases and the relative positions of it with respect to the sun led to the general agreement in very late antiquity that indeed the phases are caused by the sun's light reflected from the moon. This is the first important inference essential to the ancient measurement to be discussed here.

The planets also have sidereal periods and these were recognized as clues to the order of the cosmos. The periods of Venus and Mercury are the same as the sun's, namely one solar year. This is because these two planets always remain near the sun in the sky as either so-called morning stars or evening stars; they either rise before or with the sun in the morning or set with or after the sun in the evening. They are therefore never visible overhead at midnight, as are the other three planets. The sidereal periods of Mars, Jupiter, and Saturn form a hierarchy of 687 days, 12 years, and 30 years, respectively. These numbers came in handy for the ancient Greeks when they tried to frame a physical picture of the cosmos, something apparently the Babylonians never did. The Greeks knew that the moon was closer than the sun, since the moon passes in front of the sun during a solar eclipse. Thus they made the inference that, since the moon's sidereal period is less than the sun's, then all heavenly bodies are arranged according to the hierarchy of their periods. This meant Saturn is the farthest planet, followed closer by Jupiter and then Mars.

But then a problem arises because Venus, Mercury, and the sun have the identical sidereal periods of one solar year. How can we infer a hierarchy if the numbers are the same? In early Greek astronomical texts there is no consensus on their arrangement; one finds all possible permutations of these three bodies. Plato, it seems, placed Venus and Mercury, in that order, beyond the sun and below Mars, as did Archimedes. The Alexandrian astronomer Ptolemy in the 2nd century AD set what became the conventional arrangement carried down through the Middle Ages and Renaissance. In his *Almagest,* a work that became *the* reference of astronomers for over a millennium, he put forth this arrangement, beginning with the central Earth: the moon, Mercury, Venus, the sun, Mars, Jupiter, and Saturn. His rationale for the sun's position: "By putting the sun in the middle [of the grouping, between Venus and Mars], it is more in accordance with the nature [of the bodies] in thus separating those which reach all possible [angular] distances from the sun [namely, Mars, Jupiter, and Saturn] and those which do not do so, but always move in its vicinity [namely, Mercury and Venus]." The sun thus separated two classes of planets according to their behavior with respect to the sun. Ptolemy is surely stretching for an answer, but what was he to do when the order was not directly given by the data? Yet, there was a further problem: what were the relative positions of the two closest planets? Ptolemy placed Mercury closer to Earth than Venus, based on a fact and an assumption. The motion of Mercury in all its details is the most complex of the visible planets. Also, the moon, because of the various wobbles it makes, displays the most complex motion of all these bodies. So, because of the complexity of their motions, Ptolemy chose to keep the moon and Mercury close together, as if that were the way the Creator arranged such things, from the simplest motions near the stars to the more complex near the central Earth.

So Mercury and Venus, in that order, were between the moon and sun, with Mars, Jupiter, and Saturn extending to the sphere of the fixed stars. And that was as far as he explored such physical issues in the *Almagest,* which is primarily a treatise for predicting celestial events—not understanding them. Of course these numbers and the logic alone determined only the hierarchical arrangement of the bodies, not their distances.

2.3. Aristarchus's Measurement and Ptolemy's Model

Half a millennium before the *Almagest,* the search for celestial distances made a leap forward in the work of Aristarchus. Living in the 3rd century BC, he was sandwiched between two generations: after Euclid's synthesis of mathematics in the longest-running textbook of all time, *The Elements [of Geometry],* and before the brilliant mathematical-physicist Archimedes. In Aristarchus's only surviving work, *On the Sizes and Distances of the Sun and Moon,* he cleverly devised a way of measuring the relative distances among Earth, the moon, and the sun, based on a simple geometrical rule and knowledge that the moon's phases are due to light being reflected from the sun. He realized that when the moon is half-lit (say, in its first quarter after the new moon, called quadrature) imaginary lines from Earth to the moon and then from the moon to the sun form a right angle (Fig. 2.1). This meant that by measuring angle α, the ratio *EM/ES* could be obtained from similar triangles, that is the relative distances of those bodies. Setting *EM* to 1 unit (unity) he therefore able to measure how much farther the sun is from Earth than the moon.

In principle this is a straightforward measurement; in actuality myriad problems arise. Since the sun and moon are both moving through their sidereal periods, the measurement must not only be made on the day of the first quarter, but within about a minute or so of quadrature; otherwise the result will be grossly in error. Another source of error is clear from a consideration of the triangle. Figure 2.1 is

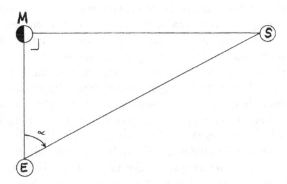

FIGURE 2.1. Aristarchus: relative distances of the sun and moon. When the angle *EMS* is a right angle (during the first or third quarters of the moon), the triangle is Pythagorean, and by measuring the visual angle α the ratio *EM/ES* follows. For α, Aristarchus got 87°, and setting *EM* = 1, he deduced *ES* = 19.

not drawn to scale; the actual triangle is very long and narrow because α is close to 90°. This means that for even minor changes in the measurement of α, just fractions of a degree, there are major differences between *EM* and *ES*. Recognizing this, as well as perhaps the lack of precision of Aristarchus's measuring instrument, we should not be surprised to find that the relative numbers he obtained were far from correct. He probably obtained a range of numbers, and it seems that he erred toward what seemed physically reasonable. The measurement he published was $\alpha = 87°$. Trigonometry was not yet invented, so he could not directly convert this to an exact number; he said the sun was 18 to 20 times farther than the moon (with *EM* = 1). This was often rounded off to 19. We assume this "reasonable" value was near the small end of the range of numbers, since he needed room for three more planets beyond the sun to fit into this geocentric cosmos. Aristarchus's value was generally accepted in late antiquity and after.

Another reason Aristarchus probably chose the smaller value follows from the next measurement he made. Recall the title of his work, *On the Sizes and Distances of the Sun and Moon*. The sizes are relative sizes (i.e., visual diameters) of the moon and sun. A dime held at arm's length covers the full moon. (Try it, if you don't believe me.) What does this tell us about the actual size of the moon? Absolutely nothing. In astronomy, as in terrestrial surveying, size and distance are coupled together. Both the full moon and the sun occlude the same angle (about ½ degree of arc) of our visual field, but their actual sizes are a function of how far away they are. That they cover (or eclipse) the same visual disk in the sky is why solar eclipses are so spectacular on Earth, since the moon just covers the sun. (There are often slight variations of this, since planetary orbits are not perfect circles.)

This amazing coincidence, from a modern viewpoint, is worth pondering briefly. There are four variables that make this so: the relative sizes of the sun and moon and their distances from Earth. Change any one of these and the moon's disk is visually either much larger or much smaller than the sun. For us, the accidental nature of this is especially remarkable when contemplating the evolution of our sun-centered solar system. But in earlier times, when it was believed that the universe came into existence all at once, such perfect fits were often seen as part of the Creator's purposeful design. This idea lingered as late as the 17th century in the mind of Johannes Kepler, who believed this arrangement was made by God so that humans may enjoy the breathtaking spectacle of solar eclipses (see section 6.2).

Figure 2.2 shows how Aristarchus used this fact to measure the relative sizes of the sun and moon. It follows from the geometry of similar triangles that their relative sizes are the same as their relative distances ($r/d = R/D$). Thus the sun must be 19 times larger than the moon, which is quite a significant difference. If he had chosen a larger number for the relative distance, say 100, then the sun would have to be that much larger than the moon. We thus conclude that Aristarchus made a realistic deduction from the data, picking what seemed physically plausible for these astronomical magnitudes, namely the smaller sizes and corresponding distances.

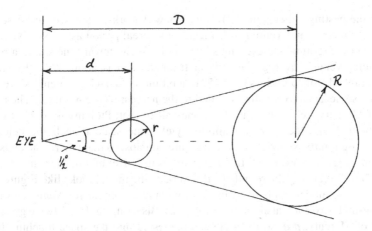

FIGURE 2.2. Aristarchus: relative sizes of the sun and moon. Since both the sun and moon occlude the same visual angle ($1/2°$) in the sky (namely, a dime held at arms' length), this diagram shows that their corresponding sizes are the same ratio as their relative distances. That is, $d/r = D/R$, and therefore $d/D = r/R$. From Figure 2.1, it follows that the sun is 19 times larger than the moon.

So how close were the measurements? Nowhere near. The sun is actually about 400 times further than the moon. The exact measurement is $\alpha = 89°$ 51 minutes, only 9 arc-minutes away from being a right angle. (As noted, a small change in angle α can make a large difference in the relative distances, since the triangle is extremely elongated.) It also means that the sun is about 400 times larger than the moon. But such an idea was almost unthinkable in antiquity. Even today, in trying to fathom ourselves as ancient astronomers and thinking in terms of a common-sense view of the possible sizes of things moving in the celestial realm, around 19 seems close to a maximum value for these measurements (at least, I think it seems to be so).

A curious thing happened when Aristarchus's measurements later got into the hands of Ptolemy. The astronomy of the *Almagest* is confined to relating observational data to a geometrical system of circles upon circles from which future observations can be predicted. It's what may be called a *mathematical* model of the cosmos. (Some of the details are in section 3.1). As a first approximation (considering the major motions of the planets only) this model fits the data and affords valid predictions from the model—or, as the ancients said, it "saves the appearances."

In a later work, a small treatise titled the *Planetary Hypotheses*, Ptolemy went beyond merely "saving appearances," to speculating on the physical nature of the heavens. Here he put forward what may be called a *physical* model—although, as the title indicates, hypothetically. Conceiving of the circles of the *Almagest* as actual physical entitles, such as wheels or spheres, he constructed a sort of scale model of the cosmos from observational data (I will call it the nesting-spheres model). Here's how it works. As noted, the first two planets (Mercury and Venus) have the same sidereal period as the sun, yet they move in the morning and evening sky away from and toward the sun in a regular cycle. Venus's average elongation from the sun is about 46°. (He knew Venus ranged from a bit less than 45° to a bit more than 47°. Ptolemy's average was the same as today's value rounded to the nearest whole number.) The average of Mercury's elongation is about one half of that; Ptolemy used 21° (today it's about 23°). One more assumption yields the scale model: the celestial spheres are tightly nested together; there are no gaps, so that where one ends the next one begins. Now, recall the hierarchy from Earth: moon, Mercury, Venus, sun. Therefore, using the rounded-off numbers, the model looks like Figure 2.3.

This scale model is obtained by first drawing the circle for Venus and then nesting-in the axes h-k at about 92° (46° × 2). Bisecting angle h-k twice gives the range of Mercury, p-q, so its circle can be nested into the space touching lines p and q and the epicycle of Venus. This, finally, specifies the circle for the moon, just touching Mercury's epicycle. Hence the circles have been drawn to scale based on the supposition of there being perfectly nesting spheres. The data are correct, the geometry is correct, and, given the nesting-spheres assumption, it portrays the relative distances of the bodies.

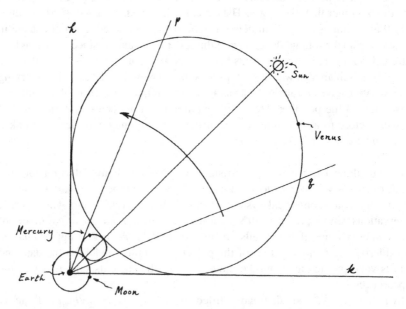

FIGURE 2.3. Ptolemy's scale model: Mercury and Venus. The epicycle of Venus is drawn to scale, from h to k (about 90°). By assuming the nesting-spheres model, the epicycle of Mercury is forced into the space between p and q (since its extent of distance from the sun is about half of Venus's), and the relative size of the moon's circle around Earth then follows. Using a similar construction, Ptolemy deduced that the sun is 19 times further from Earth than the moon.

Setting *EM* for the radius of the Moon's orbit, and *ES* for the Earth to sun distance, one can measure the ratio *EM/ES* from the diagram. The larger the drawing the more accurate it is, but using a standard $8^{1}/_{2}'' \times 11''$ sheet of paper, with the diameter of Venus's epicycle of about 17 cm, I obtained *EM* = 1.4 cm and *ES* = 20.6 cm. Setting *EM* to 1 gives *ES* = 14.7 or about 15. Today we might call this a ballpark figure. My reconstruction of Ptolemy's measurement of the ratio *EM/ES* is a simplified version of Ptolemy's more complex method, since it appears that he never made an actual scale drawing. Using instead a computational method, he obtained a range of values from 18 to $18^{1}/_{2}$.

This brings us to the crux of this story. All these values, especially Ptolemy's, are in the same range of that derived earlier by Aristarchus using an entirely different method. Both yield the same approximate number 19! For the late antique world this was a clear-cut case of the "convergence of data" and is thus further evidence for the correctness of both Aristarchus's measurement and Ptolemy's model. After all, if we measure the same thing two different ways and get about the same number, it's reasonable to assume that the two measurements reinforce each other because they are both correct. Just as scientists measuring the speed of light by different methods were sure that they were converging toward the more precise value, so ancient astronomers were convinced that the 1/19 ratio was near the real value.

What is reasonable now was reasonable then. Whether in ancient times or today, the logic is the same. Yet for us this ancient result is astonishing, since we know that it is nowhere near the real ratio of about 1/400. It was a mere historical accident—a coincidental concurrence—that these two measurements converged.

Beyond the moral of this story—that matching data do not necessarily converge toward reality—there is an irony too. It is not surprising that throughout the Middle Ages and Renaissance—namely, for about a millennium and a half after Ptolemy—the convergence of these measurements was a potent argument not only for the 1/19 ratio but also that the geocentric model of the universe was indeed correct—namely, the one made by God. Nevertheless, the name Aristarchus also conjures up another astronomical model of considerable import: the heliocentric (or sun-centered) system. As far as we know he was the first to put forward the notion that Earth moves around the sun. (He was not the first to conceive of a moving Earth, however; Pythagoras did that about 250 years earlier, but his Earth, along with the sun, moved around a fire centered on the sphere of the fixed stars.) Unfortunately, we do not have Aristarchus's words on this, since the treatise in which it appeared is lost. We only know about it from what Archimedes later reports in his work, *The Sand-Reckoner.* There he says that Aristarchus put forward the hypothesis that Earth moves in a circular orbit about the sun and that the sphere of the fixed stars is extremely large. The reason for the later assertion is this: if Earth really moves, then it follows that every half-year there should be a visible shift in the positions of the stars (called stellar parallax).

In Figure 2.4 notice that this angular shift should be 2α. But no such parallax is observed; therefore, in order still to hold to a moving Earth, Earth's orbit must be

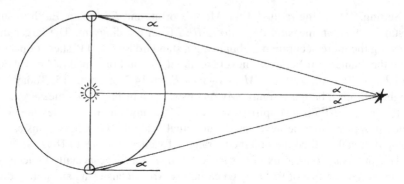

FIGURE 2.4. Stellar parallax. A star seen from a moving Earth should visually shift in the sky by twice angle α every half-year.

extremely small compared to the distance of the stars, so small that the parallax is not perceptible. Said otherwise, in Figure 2.4, angle α is so small as to be unde-tectable. Needless to say, and on the contrary, over the centuries the lack of stellar parallax was almost always interpreted as proof that Earth did *not* move, rather than that the stars were an enormous distance from Earth and the sun (see section 5.1).

In the end Aristarchus's name was persistently associated with the idea, how-ever wrong it might have been thought to be, of a heliocentric cosmos, while at the same time his measurement of the Earth–moon–sun distance reinforced the geocentric system. Logically the irony only holds if the measurement were neces-sarily coupled with heliocentrism, but it is not. Instead, the measurement alone is based only on the relative positions of the three bodies and hence is independent of the system used. It works for either the geocentric or the heliocentric model, or any other compatible one.

Notes and References

I have used Galileo's *Two New Sciences,* translated by Stillman Drake (Toronto: Wall & Thompson, 1989), quotation on pp. 50–51.

Ptolemy's Almagest, translated by G.J. Toomer (Berlin: Springer-Verlag, 1984), quotation from pp. 419–420. An English translation of Aristarchus's *On the Sizes and Distances of the Sun and Moon* is in Thomas Heath, *Aristarchus of Samos: The Ancient Copernican* (Oxford: Clarendon Press, 1966), pp. 351–414.

Albert Van Helden, *Measuring the Universe: Cosmic Dimensions from Aristarchus to Halley* (Chicago: University of Chicago Press, 1985), pp. 4–40. Otto Neugebauer, *A History of Ancient Mathematical Astronomy* (Berlin: Springer-Verlag, 1975), pp. 111–112. Bernard R. Goldstein, "The Arabic Version of Ptolemy's *Planetary Hypotheses,"* *Transactions of the American Philosophical Society,* 57, part 4 (1967).

3
The Rationality of Simplicity:
Copernicus on Planetary Motion

A prevalent textbook view of scientific change often appeals to the role of anomalies: contradictions that sometimes arise between new empirical data and the conventional model at the time. One would think, therefore, that when Copernicus put forward the radical idea that Earth really moves around the sun, in contradiction to our experience of a fixed Earth, he would have been forced to do so by some anomaly in the astronomical data. In fact, however, there was no pressing anomaly; there was nothing new or troubling under the sun or in the night sky in the early 16th century. Furthermore, and contrary to what some books say (including, as we will see, Copernicus himself!), the geocentric model was working well, perhaps as good as it ever did, having been fine-tuned by Renaissance astronomers. So how could Copernicus justify such an extreme idea, rejecting thousands of years of common sense? This chapter provides an answer.

3.1. Planetary Motion: Geocentrism

The idea of a heliocentric (sun-centered) universe was first put forward in the third century BC by Aristarchus (see section 2.3). Although rejected at the time, and for good reasons, the idea was never forgotten. From late antiquity through the Renaissance, his name was associated with this curious but obviously wrong model of the heavens that entails a moving Earth. When Copernicus in the 16th century seriously put forward the heliocentric model, he was, in one sense, merely proposing a revival (renaissance) of Aristarchus's model. *Why* Copernicus did so remains an unanswered question today, despite decades of work on his life as well as numerous attempts to reconstruct his thought processes. (There are various hypothetical reconstructions, but still no consensus among historians of science.) But we do have Copernicus's *justification* for believing the model to be true. Essentially it was based on the motions of the planets, and his solution to what probably was seen as a puzzle when first observed by ancient astronomers.

The geocentric viewpoint, today and in prehistoric times, entails the following behavior for the planets. As do the sun and moon, the planets slowly move through the background of the stars from west to east (the cycle being their

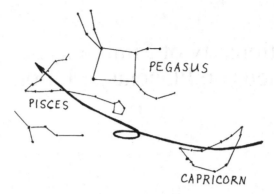

FIGURE 3.1. Retrograde motion. Planets, as they cycle from west to east through the backdrop of the stars, periodically reverse this motion, making a loop in the sky.

sidereal periods), at the same time as they daily rise and set from east to west (see section 2.2). For the sun this motion generates the solar year. But the planets do more: as they move through the stars from west to east they also periodically *reverse* this motion (called retrograde motion by the ancients), producing periodic loops through the sky (Fig. 3.1). This is why the Greeks called them "planets," that is, wanderers. The repeatable cycle of time between loops is called a synodic period. In addition, during this retrograde motion, the planets become brighter, indicating that they are closer to Earth. (These two cycles, sidereal and synodic, comprise the major motions of the planets on the geocentric system; we are ignoring various other minor shifts.)

The sidereal periods through the stars and the accompanying synodic periods are unique for each planet (see the first two columns in Table 3.1). However, they are not entirely arbitrary, for there is a singular and nonarbitrary parameter among the planets linking these motions. To see this we need to construct the mathematical model for the planets used in the second century AD by Ptolemy in his *Almagest*. Ptolemy used an epicyclical system—attributed to Apollonius in the third century BC—where two circles suffice to generate the major observed motions of a planet.

In Figure 3.2 the large circle is called the deferent, and the small circle the epicycle; the planet is attached to the epicycle and hence the combination of the two continuously rotating motions produces the required loops. Note how this model at once accounts for retrograde motion and the corresponding change of brightness. To view an animation of this and other models in motion see the Web site of Dennis Duke cited in the notes at the end of the chapter.

Considering these major motions, parameters in the model are quantitatively correlated with two sets of observational data of the planets: the times from west to east through the stars (the sidereal periods, T_{sid}) and the times between retrograde motions (the synodic periods, T_{syn}). Transforming, or correlating, these motions to the model, it is obvious that the observed sidereal period is simply the

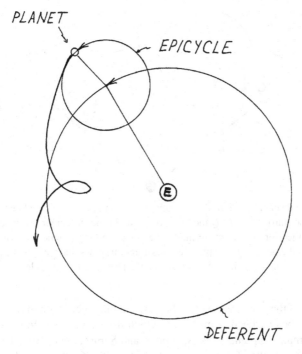

FIGURE 3.2. Epicyclical motion: deferent and epicycle. A model for retrograde motion involves two circles: a larger deferent and smaller epicycle, with the planet attached to the latter. Combining these two rotational motions produces the requisite retrograde loop of the planet (see Fig 3.1) from the point of the view of a central Earth.

period of rotation of the model's deferent. Correlating the retrograde (synodic) period to the model may seem trivial, too; merely correlate the synodic period with the period of rotation of the epicycle—but that is, in fact, wrong. To see why, look closely at Figure 3.3, and the planet's motion from P to Q to R. Notice that the time for epicycle to make one rotation around its own center (from P to Q) is less than the time required to complete a retrograde loop (P to R). The time from P to Q (not R) is the epicycle's period because the lines from the center of the epicycle to the planet are parallel. Note also that the way I have drawn this example applies primarily to Saturn and Jupiter (as the keen reader may surmise from Table 3.1). Conveniently, and importantly, there is a simple formula for calculating an epicycle's period (T_{epi}) from the data: $T_{epi} = (T_{syn} \times T_{sid})/(T_{syn} + T_{sid})$. (It is an interesting geometrical problem to derive this formula, which I leave for the inquisitive readers.) Using it, results in the numbers in the third column in Table 3.1. Note that the parameter of one solar year appears throughout the table. This is the nonarbitrary link I mentioned at the start. Why is it there?

Let's begin with Mercury and Venus, where the reason is quite obvious (see section 2.3). Since both planets are always only visible in either the morning or evening, they do not stray far from the sun, and hence their sidereal periods must

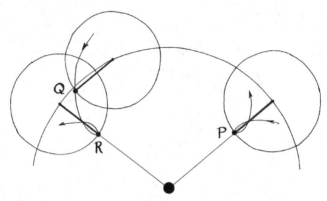

FIGURE 3.3. Epicyclical motion: synodic and epicycle periods. As the planet moves from *P* to *R*, it passes through *Q*. At *Q* the lines to *P* and *Q* from the centers of the epicycles are parallel, and therefore the epicycle made one complete loop around its own center from *P* to *Q*. This is the period of the epicycle. This diagram shows that the synodic period, from *P* to *R*, differs from, and is in fact greater than, the period of the epicycle.

be the same as the sun's, for even it they drifted only a bit, then in time they eventually would be overhead at midnight, which they never are. On the other hand, when the remaining planets, Mars, Jupiter, and Saturn, are overhead at midnight, they are in the middle of their retrograde motion: that is, they are closest to Earth and at their brightest intensity.

Perhaps I should be more precise about that phrase "overhead at midnight." There is an important imaginary line of sight called the meridian: starting from due north on the horizon passing through the zenith (the point directly overhead) and terminating directly south to the horizon, this line slices the sky in half. When we face south, celestial objects arc across the sky during the night from east to west, reaching their highest point when crossing the meridian. Therefore, when a planet is in "opposition," that is, the sun is exactly opposite on the other side of Earth, it is obviously midnight, and if the planet crosses the meridian at this time, then it is in the middle of its retrograde loop and also closest to Earth. Using Ptolemy's model, at opposition a line from the center of the epicycle to the planet would continue through Earth and meet the sun on the

TABLE 3.1. Ptolemy's system.

Planet	T_{sid}	T_{syn}	T_{epi}
Mercury	1 year	116 days	88 days
Venus	1 year	584 days	225 days
Sun	1 year	—	—
Mars	687 days	780 days	1 year
Jupiter	12 years	399 days	1 year
Saturn	30 years	378 days	1 year

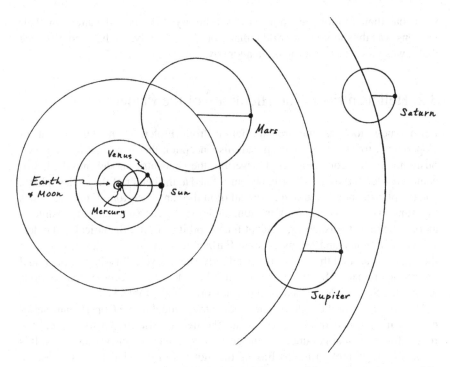

FIGURE 3.4. Ptolemy's scale model. This is a scaled diagram of the nesting-spheres model. Note that the lines from the centers of the epicycles to the planets Mars, Jupiter, and Saturn are all parallel to the line from Earth to the sun. This accounts for the correlation of these planetary motions with that of the sun; namely, at opposition they cross the meridian at midnight. In contrast, the centers of the epicycles of Mercury and Venus are on a line with the sun (see Fig 2.3) since they always are only visible in the morning or evening, never straying far from the sun.

other side. This means, furthermore, that for all three planets these lines are *always* parallel to a line from Earth to the sun (as in Fig 3.4). This corresponds to the fact that the periods of the epicycles of Mars, Jupiter, and Saturn are all the same 1 year. This is important, as shall be seen, especially when converting to the heliocentric model. Nonetheless, and unfortunately, I have seen purportedly Ptolemaic models drawn incorrectly in textbooks, that is, without these lines being drawn parallel.

I should also point out that when facing south and a star or planet reaches the top of its arc at the meridian, this is another way to find the meridian—that is, due south. As well, when facing north, the stars arc the other way, reaching their lowest point at the meridian, and likewise this specifies due north. From these two directions, due east and west can likewise be reckoned—all without a compass.

The 1-year parameter or link among the planets was recognized over the ages, as far as we know, as merely that—a common parameter connecting the planets, such that their motions are not entirely arbitrary. The overall pattern of their motions was thus more system-like than not. Significantly, we have no evidence that it was viewed as a puzzle or an anomaly.

3.2. Heliocentrism: The Hierarchy of the Planets

Let us now look at the transformation from Ptolemy's model to that of Copernicus, for there is a widespread misconception that is related to the previously noted erroneous diagram. Converting the geocentric model into the heliocentric model, the sun, not Earth, becomes fixed at the center of the stellar sphere, and the planets, now including Earth, all orbit the sun. Therefore, Earth subsumes the sun's yearly (geocentric) motion, and the 1-year (geocentric) parameter among the planets disappears, except for the orbit of Earth. Specifically, the deferents of Mercury and Venus become Earth's year, and their epicycles become their periods around the sun; correspondingly, the epicycles of Mars, Jupiter, and Saturn are replaced by Earth's year, and their deferents become their periods around the Sun. Table 3.1 is therefore transformed into Table 3.2.

Having transferred the Ptolemaic (geocentric) model to the Copernican (heliocentric) model, and moreover believing the heliocentric model to be true, the reason for the 1-year parameter in the geocentric model becomes transparent. It is Earth's (yearly) motion that is linking the motions of the planets as we observe them from a seemingly fixed position. Said another (geometrical) way, the retrograde motions of Mars, Jupiter, and Saturn are due to the relative motions of the planets and a moving Earth (Fig. 3.5). Their apparent backing-up is an illusion, as Earth, with the faster orbit, passes a planet. (This may be one of the first cases of what is later important for Galileo and Einstein, namely, the relativity of motion.) And Mercury and Venus are always observationally near the sun in the geocentric model because they really do orbit the sun between it and us. What was geocentrically a common link among the planets becomes a clue to the structure of the universe. Is this heliocentric model, therefore, not more system-like? Is this not a simpler explanation of the planetary motions? In retrospect, was not the 1-year parameter really an anomaly in the geocentric model? Affirmative answers to these questions are alone very compelling reasons for the authenticity of the Copernican model.

TABLE 3.2. Copernicus's system.

Planet	Period
Mercury	88 days
Venus	225 days
Earth	1 year
Mars	687 days
Jupiter	12 years
Saturn	30 years

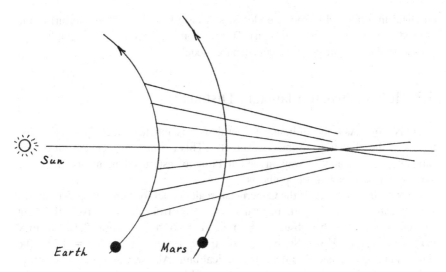

FIGURE 3.5. Retrograde illusion. Assuming the heliocentric model, retrograde motion is understood as an illusion caused by the relative motion of Earth and the planet. In this case, Mars appears to back up through the stars as Earth passes it.

This brings me to a frequent misconception, or myth. It is often noted, correctly, that both Ptolemy's and Copernicus's systems are observationally equivalent: this is obviously true, otherwise each model would not work! But this fact is then sometimes generalized to positing that geocentrism and heliocentrism are equivalent, that they are the same geometrically or mathematically or even exactly the same; in a phrase, there is an isomorphism between them. Such assertions are often found within the context of arguing for the relativity of scientific explanation, as if replacing Ptolemy with Copernicus was a mere change of perspective or viewpoint (or a paradigm shift, in the trendy vernacular). But this is patently false. Only *some*-types of Earth-centered deferent-epicycle systems can be transformed into Copernican-type sun-centered systems. In fact, only geocentric systems that have parallel lines as shown in Figure 3.4 can be transformed into a sun-centered system. For all others it is impossible. (See section 5.2 on the Tychonic system.) Said another way, Ptolemy's system may be transformed into a heliocentric one because our system really is heliocentric! That is why there is the 1-year parameter in Ptolemy's model. As noted, in retrospect, this was as clue that heliocentrism should have been taken seriously. Alas, it took a long time for astronomers to realize this.

Comparing Tables 3.1 and 3.2 also expose other ways Copernicus's model is simpler than Ptolemy's. Where Ptolemy required two circles (deferent and epicycle) for the major motions of the planets, Copernicus used only one for the same motions, because Earth's motion, so to speak, accounted for the difference. Also the resulting single column of Copernicus casts light upon (or solves, if you believe him) an ancient problem—namely the hierarchy of the planets (see section 2.2). Notice how there no longer is an option for the planet closet to the center (here the sun); all are arranged in a hierarchy from the closest and fastest (with an 88-day orbit) to the farthest and slowest (a 30-year orbit). This is another compelling (and simplifying) reason for the acceptance of the heliocentric model.

3.3. Heliocentrism: Planetary Distances

There is more. As a sort of bonus, Copernicus was able to deduce the relative distances of the planets from the sun, directly. This is important, because it differs fundamentally from the way in which Ptolemy estimated planetary distances using the nesting-spheres hypothesis.

Look at the geometry of the Copernican scale model for Venus (Fig. 3.6, drawn when Venus is at its maximum visual elongation from the sun). Note the right triangle *EVS*; since by observation angle $\alpha = 46°$ (see section 2.3), we may derive the ratio *SV/SE*. Setting *SE* to unity, results in the *SV* = 0.72. The Earth–sun distance is called the astronomical unit (AU), which is customarily set to 1. A similar calculation gives the distance of Mercury as 0.36. Thus between the

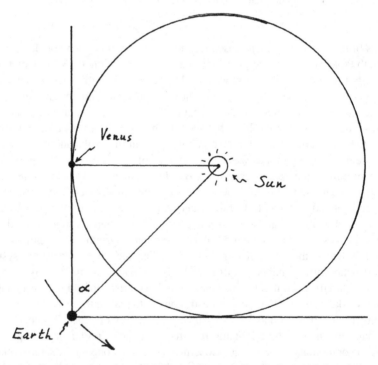

FIGURE 3.6. Copernicus's scale model: Venus. On the heliocentric model the relative scale of Venus is the same as on the geocentric model (compare Fig. 2.3); however, the sun is placed at the center of the orbit of Venus, and of course, Earth is moving around the sun. The planet's epicycle on the geocentric model is now its orbit around the sun.

sun and Earth, Mercury is about one-third the distance and Venus about three-fourths the distance. The method for deriving the planetary distances of Mars, Jupiter, and Saturn is more complex (but a fine exercise in geometry and algebra, for the inquisitive reader). Doing so result in all the following relative distances: Mercury (0.36), Venus (0.72), Earth (1), Mars (1.5) Jupiter (5), and Saturn (9). The hierarchy of these distances parallels the hierarchy of their periods: 88 days, 225 days, 1 year, 687 days, 12 years, and 30 years, respectively. In the first chapter of *De Revolutionibus* (1543), Copernicus waxes poetic on the aesthetics of these orders: "In this arrangement . . .we discover a marvelous symmetry of the universe, and an established harmonious linkage between the motion of the spheres [that is, the period of the planet] and their size [that is, the relative distance], such as can be found in no other way." He is right (well almost; see section 5.2). There truly is a remarkable parallel between these two sequences of numbers.

It is important, too, that they are derived directly from the geometrical structure of the heliocentric model, without introducing any assumptions. This is in contrast to the (additional and ad hoc) assumption of nesting spheres in Ptolemy's derivation of distances. Finally, the clincher is this: the sequence of distances that follow from Copernicus's model were not accessible on Ptolemy's model, because there is no right triangle among Earth, the sun, and the planets from which to deduce corresponding ratios (compare Figures 2.3 and 3.6; note that in Fig. 2.3 there is no position of Venus where there is a right angle between it and Earth and the sun). From the Pythagorean (right) triangles in Copernicus's model the planetary distances follow *directly*. The only assumption, of course, is the heliocentric model.

There is another contrast between the models that is not well known. Computing the distances to Saturn in terms of Earth–sun distances, Ptolemy obtained a range of 12 to 17 AU, for an average value of about 14 AU. (Ptolemy really used the radius of Earth as a unit on his geocentric model, but I am converting this to AU for comparison with Copernicus's data.) It is of more than passing interest to note that Copernicus's calculation of the distance to the last planet, Saturn (9 AU), is much *less* than the distance arrived at by Ptolemy's nesting-spheres hypothesis. This is generally not known; indeed it is often said that the old geocentric system was smaller and more compact than the heliocentric model.

But Ptolemy's system is larger only for the planets. Since the celestial sphere is nested right after Saturn's, the distance to the stars is thus about 14 AU. Now recall that Archimedes mentioned that Aristarchus, in putting forth the heliocentric model, also affirmed that the sphere of the fixed stars is extremely large. The reason was the absence of a visible stellar parallax that should be perceived if Earth moved (see Fig. 2.4). So Copernicus, in order to account for the absence of a semiannual parallax, says that the space between Saturn and the stars is "vast." When we include the stars, then Copernicus's cosmos is much, much larger than Ptolemy's.

3.4. Copernicus and Simplicity

Returning to the singular, aesthetic unity—the fundamental simplicity—of the Copernican system, the two hierarchies (planetary periods and distances) impart a distinct oneness to Copernicus's model. In contrast, this is not true for Ptolemy's model; as seen, each planet's motions remained very nearly distinct, so it is not really a system after all. Copernicus recognized this contrast and pounced on it right at the start of his book. In his preface he asserts that the aim of astronomy is to "deduce . . .the structure of the universe and the true symmetry of its parts," and he contrasted this with the methodology of those who use the Ptolemaic model, and, interestingly enough, he employed a metaphor borrowed from art theory. The Ptolemaic method, he writes, is "just like someone [making a drawing of a person and] taking from various places hands, feet, a head, and other pieces, very well depicted, it may be, but not for the representation of a single person; since these fragments would not belong to one another at all, a monster rather than a man would be put together from them." Ancient astronomy had produced a monster, but Copernicus slew it by deducing "the true symmetry of its parts." Listen to this definition of beauty from Leon Battista Alberti, a key theorist of the early Renaissance, taken from his treatise "On Architecture" (1452): "I shall define Beauty to be a harmony of all the parts . . .fitted together with such proportion and connection, that nothing could be added, diminished or altered, but for the worse." Now let me quote Copernicus, again: "In this arrangement . . .we discover a marvelous symmetry of the universe, and an established harmonious linkage between the motion of the spheres and their size, such as can be found in no other way." Both are what may be called a holistic viewpoint. Such an aesthetic maxim Alberti also applied to the depiction of the human body, for in his earlier treatise "On Painting" (1435) he called for the parts of the body to "go well together" so as to "correspond to a single beauty." The underlying objective was not without foundation; it was common practice for novice Renaissance artists to be taught to draw by depicting individual body parts, such as plaster casts of feet or arms (often based on classical sculpture). Giorgio Vasari, the Renaissance biographer of artists, and Michelangelo's friend, speaks of achieving "the greatest possible beauty" by "joining together these most beautiful things, hands, heads, bodies, and legs."

Indeed Copernicus (1473–1543) lived in the age of the High Renaissance giants in art: Leonardo died in 1519, Raphael in 1520, and Michelangelo in 1564. Copernicus had spent the years 1496 to 1503 in Italy pursuing postsecondary degrees in law, medicine (which entailed anatomy), and the liberal arts (including astronomy and mathematics). There is also evidence that he was an amateur painter. It is not surprising, therefore, to see him borrowing a principle of unity from art theory to bolster his astronomical model.

Copernicus and Michelangelo

Another connection between Copernicus and art has been put forward by Valerie Shrimplin-Evangelidis. Her thesis is this: the composition of Michelangelo's *Last Judgment* fresco on the altar wall of the Sistine Chapel reflects the heliocentric cosmos of Copernicus.

She asserts that Copernicus and Michelangelo were influenced by various ideologies of the Renaissance, especially neoplatonism, with its emphasis on sun symbolism and the analogy of the Deity with the sun. The idea appears in the text of Copernicus's *De Revolutionibus* (1543), in a famous passage comparing the sun with a god, although the later Copernican scholar Edward Rosen has convinced me that this has nothing to do with neoplatonism. Copernicus also put forward a brief sketch of the idea of heliocentrism much earlier, in a private work, the *Commentariolus*. The date of composition is unknown, but numerous handwritten copies of it were circulating among students of astronomy from about 1514.

In 1533 Michelangelo received his commission for the fresco from Pope Clement VII, who was knowledgeable of the Copernican system, which we know was being discussed at the time among Vatican philosophers and astronomers. The following year he died, being replaced by Pope Paul III to whom Copernicus dedicated *De Revolutionibus*. Michelangelo began the first cartoons (that is, full-scale drawings) for the fresco in the autumn of 1535 and completed it in November 1541. The source of heliocentrism, therefore, could only be the *Commentariolus*, not *De Revolutionibus*.

Although Michelangelo's composition of the *Last Judgment* comprises the traditional right–left = good–evil and up–down = heaven–hell symmetry, these dualisms are overridden by the more dominant *circular* motif, with Jesus placed at the center. This circular composition, untypical for the Last Judgment theme, Shrimplin-Evangelidis believes reflects Michelangelo's allegiance or reference to the heliocentric cosmos. What I find to be the most convincing evidence for her thesis is the yellow-golden halo of light behind Jesus and Mary near the center of the circles.

3.5. Copernicus and Complexity

Having made a case for Copernicus's model based on its geometrical and mathematical simplicity vis-à-vis Ptolemy's, I now contrarily gum up the argument by pointing to some of the complexities in Copernicus's model. As I hope to show, in the context of the times it was not as simple as it may appear today, which was at least one reason why it did not catch on quickly.

It is true that ultimately Copernicus's work on the heliocentric system formed the basis of a revolution in the history of astronomy, and in this sense the word *revolution* in the title of his treatise, *On the Revolutions,* is appropriate. Copernicus, however, used the term literally: the work was about how the spheres *revolve*. Despite its modernity in positing the motion of Earth, Copernicus's *De Revolutionibus* is conceptually steeped in the ancient world, which was not uncommon in Renaissance treatises. The work is organized in a format parallel to Ptolemy's *Almagest*. Moreover, he conceived of the planets as being attached to rotating spheres in their motions around the sun. This means that since Earth is ever tilted to the plane of its orbit, then a problem arises within the model as our planet orbits the sun.

To see this problem, begin with the geocentric system (see Fig. 4.5). The fixed stars appear as if attached to a sphere rotating daily around an axis at the pole star (Polaris, near *N*) and accordingly there is an imaginary celestial equator around this sphere 90° from Polaris. The sun's yearly motion from west to east (its sidereal period), traces out a line in the heavens called by the ancients the ecliptic (since eclipse take place about it). The ecliptic, moreover, is tilted 23¹/₂° to the celestial equator. All this is observationally true, even today, from the viewpoint of a fixed Earth. When this datum is transferred to the heliocentric model, the daily motion of Earth on its axis accounts for the daily rotation of the stellar sphere (which is now fixed), and the annual solar year becomes the period of Earth revolving around the sun, with the important stricture that Earth's axis accordingly must be tilted 23¹/₂° to a perpendicular to the plane of its orbit in order to account for the (geocentric) tilt in the ecliptic. Significantly, Earth must remain fixed in this position throughout its annual motion (Fig. 3.7).

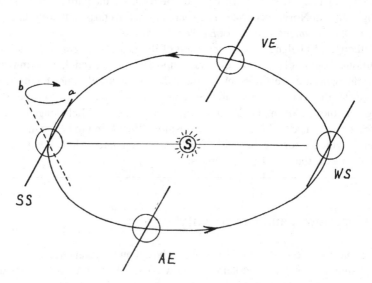

FIGURE 3.7. The tilt of the earth: Copernicus. To account for the (23¹/₂°) tilt of the ecliptic to the stellar equator when transferring from the geocentric to the heliocentric model, Copernicus tilted Earth by the same amount to the plane of motion around the Sun. Four positions are marked: summer and winter solstices, and vernal and autumnal equinoxes. Earth remains in this fixed tilted position, except for the very slow precessional motion that is shown in the diagram by the conical loop (*a* to *b*).

This is a necessary consequence of the model fitting the observational data. But a problem arises if, as is Copernicus's conception, a revolving sphere is the source of this motion; then the tilt would have to change constantly (it may help to visualize this by thinking of the sun as the hub of a wheel); otherwise, for example, beginning at the winter solstice (ws), in half a year at the summer solstice (ss) Earth would be at position b (tilted $23^1/2°$ the wrong way), whereas it should be at position a. Therefore, to preserve the tilt of Earth, Copernicus was forced to introduce a third motion (in addition to its daily and annual motions). Earth maintains its requisite tilt only if it also makes an annual *conical* motion in the opposite direction to its revolution around the sun (see the arrow at the summer solstice in Fig. 3.7). It is recommended that the reader physically test this by using one's arm held outward horizontally and hold in one's fist, say, a pen or pencil at an angle. Rotate the fist clockwise as the arm moves counterclockwise around the body; in order to keep the tilt fixed, the fist must perform this clockwise conical motion with the same period as the rotating arm.

At first this extra motion appears cumbersome and arbitrary and seemingly complicates an otherwise simple model. True enough. But it was fortuitous, too, because there was one more celestial motion, not mentioned so far, that Copernicus also had to account for heliocentrically. This is a very slow conical motion of the stellar sphere (with a 26,000-year period) called the precession of the equinoxes (for details, see section 4.4). From a heliocentric viewpoint, with the stars absolutely fixed, Earth then must assume this slow conical wobble otherwise ascribed to the stellar sphere; in other words, the motion illustrated in Fig. 3.7 for Earth at the summer solstice is precisely the motion it must perform to account for the precession of the equinoxes. Since Copernicus, as just seen, imparted such a conical (third) motion to Earth because of the physicality of the spheres implied in his model, then he simply had to add a tiny bit of slippage to this annual conical motion to give Earth one extra rotation (namely one every 26,000 years—tiny indeed) and thus accounted for the precession phenomenon. In fact, he deemed this as being so important that he devoted a major section of his book to it.

Lastly another complexity entailed in heliocentrism was forced upon him because of his resoluteness to the ancient principle of circularity. Without getting too bogged down in details, the problem is plainly seen in reverse chronology: the planetary orbits we know, since Kepler, are elliptical. Thus only for the major motions (as discussed so far) will the use of circles (deferents and epicycles) suffice to fit the data. Nevertheless, to account for *all* the data, the major and various minor motions, Copernicus was required to extend the model further. To understand his solution we must first review how Ptolemy dealt with the problem.

Obviously the same problem arose in ancient astronomy in the geocentric context. Thus, for example, it was known by Hipparchus in second century BC that the sun appears to change its speed over the course of the year (for example, the time from the vernal to the autumnal equinox is about a week less than that from the autumnal back to the vernal), and he explained this apparent contradiction to the canon of in uniform circular motion by placing the center of the sun's circle off-center (eccentric) to Earth. Similar departures of uniform rotation among the

planets led to their deferents also being placed eccentric to Earth. Even this
fiddling with the model, however, was not sufficient to fit all the data of the
planets, so smaller epicycles (call them epicyclets, since they account for
the minor motions, not retrograde motion) were thus introduced, and these further
fine-tuned the model in accommodating the data to the model—almost, but not
quite. Having used eccentrics and epicyclets, and seemingly exhausted the flexibil-
ity of the model, Ptolemy made one more change—perhaps in desperation. He
separated the center of (uniform) motion and the (eccentric) center of the circle,
thus splitting for the first time two centers that had been assumed to be coincidental (Fig. 3.8). Now there were three points of reference: Earth, the eccentric

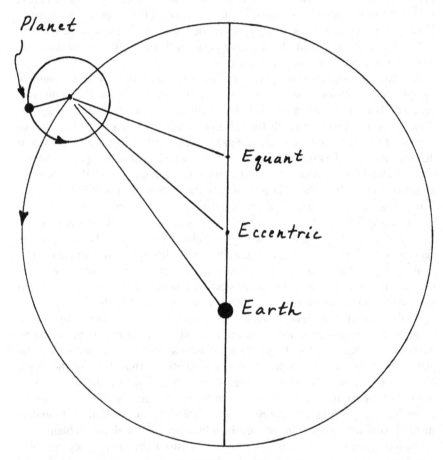

FIGURE 3.8. Ptolemy's equant. The distance between the equant point and the center of the
deferent (the eccentric point) equals the distance between the same (eccentric) point and
Earth. The center of the planet's epicycle moves with uniform rotational motion only
around the equant. Accordingly, rotations of the epicycle's center around Earth and around
the eccentric point are nonuniform. The planet's rotation on its epicycle, however, is
uniform.

(the center of the deferent), and the new point that Ptolemy called the equant (the center of uniform rotational motion). Only from the point of view of the equant is there uniform rotation with respect to the deferent. (The planet, incidentally, still is uniformly rotating about its epicycle.) The reason the equant worked—and it did—was not clear until Kepler (see section 6.3), but in the meantime astronomers were often uncomfortable or even perplexed by the equant, yet they were forced to use it to make the data and the model fit perfectly.

By Copernicus's time, the equant was taken for granted as an integral part of Ptolemaic astronomy, but he was not convinced. There is some evidence that Copernicus's quest for an alternative cosmology was motivated by a desire to rid astronomy of the equant. By switching to the heliocentric model the equant could be eliminated and still fit the data; nevertheless, he still needed eccentrics and epicyclets—but no epicycles (since retrograde motion is an illusion) or equants. (There is a caveat here: see the last paragraph in section 6.3.) In this sense, his model was not as complex as Ptolemy's, which was another reason for supporting it.

Initially this rejection of the equant may be viewed as a refutation of the ancient system, thus betraying an element of modernity in Copernicus's world-view. Maybe, but really not so. Here is his reasoning in the *Commentariolus*. After observing that Ptolemy's model fits the data, he mentions the equant as a "difficulty." This makes Ptolemy's system "neither sufficiently absolute [a phase whose meaning is not immediately clear] nor sufficiently pleasing to the mind [this now-famous phrase is clearly directed toward the aesthetic nature, or lack thereof, of a model]." The meaning of the former phase is explained, I think, subsequently. He goes on to say that he tried to correct the "defects" in Ptolemy's system by a "more reasonable arrangement of circles" in which "everything would move uniformly about its proper center [namely, the center of the circle, not the equant], as the rule of *absolute motion* [my italics] requires." This "arrangement," of course, was heliocentrism. The motivation, there is little doubt, was therefore to eliminate the equant and return the model to its unspoiled "absolute" beauty—namely, all motion being in circles, and all motion being uniformly about their geometrical centers; that's what he means by the phrase "sufficiently absolute." Thus it was not just a step forward, beyond Ptolemy, but more so a step backward to the initial Greek aesthetic framework. Hence the meaning of the phrase "everything would move uniformly about its proper center, as the rule of absolute motion requires" is evident. Spoken like a true Renaissance man, in the original sense of the term. This is—if I were asked to speculate—the reason Copernicus moved the sun. And speaking of the Renaissance, do not forget that heliocentrism itself was an old idea, first put forward, as far as Copernicus knew, by Aristarchus about 500 years before Ptolemy, and now retrieved and reasserted by Copernicus.

Copernicus's model thus was simpler than Ptolemy's, but complex nonetheless. Yet not as simple as his one big-picture diagram in *De Revolutionibus* (Fig. 3.9), illustrating the heliocentric model from the central sun, through the planets (including Earth) to the stars. This famous diagram, to be sure, is neither complete (since it only includes the major circles) nor (obviously not)

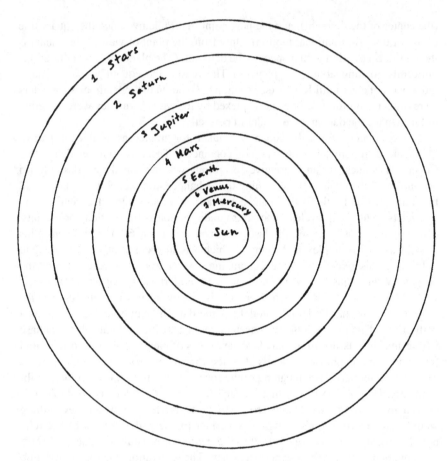

FIGURE 3.9. Copernicus: heliocentric diagram. A sketch of Copernicus's diagram of the heliocentric system, based on his manuscript for *De Revolutionibus*. Note that it is a conceptual diagram; that is, it is not drawn to scale.

to scale. It is what we may call a conceptual diagram of the system. The first detailed and scaled diagram would not be drawn until the work of Kepler (for example, Fig. 5.2).

Notes and References

I have used G.J. Toomer's translation of *Ptolemy's Almagest* (New York & Berlin: Springer-Verlag, 1984). The correlation of the epicycle periods of Mars, Jupiter, and Saturn with the sun is found on pp. 424–425 and 480–484, although Ptolemy's "formula" is in a different format than mine. I wish to thank Dennis Duke (Physics, Florida State University) for these references in Ptolemy. His superb Web site displays animated versions of various astronomical systems: http://www.csit.fsu.edu/~dduke/models.

I have used Edward Rosen's translation of Copernicus's *On the Revolutions* (Baltimore: Johns Hopkins University Press, 1978), quotations from pages 4 and 22. Copernicus's *Commentariolus* is in *Three Copernican Treatises,* translated by Edward Rosen (New York; Dover, 1959), quotations from pp. 57–58.

The myth of the equivalence of Ptolemy and Copernicus is clearly presented in, Keith Hutchison, "Sunspots, Galileo, and the Orbit of the Earth," *Isis,* 81 (1990), pp. 68–74.

On the nonanomalous view of the motion of the planets before Copernicus, as well as an insightful discussion of anomaly itself, see Alan Lightman and Owen Gingerich, "When Do Anomalies Begin?," *Science,* 255 (Feb. 7, 1991), pp. 690–695.

Years ago I noticed the striking similarly between Copernicus's concept of astronomical aesthetics and Alberti's definition of beauty and decided to explore it someday; and there it sat on the backburner of my research endeavors. In the meantime someone else (Jeroen Stumpel) also spotted the similarity and, I am pleased to say, pursued the research, producing the article from which I have drawn most of my information; see his, "On Painting and Planets: A Note on Art Theory and the Copernican Revolution," in *Three Cultures: Fifteen Lectures on the Confrontation of Academic Cultures* (The Hague: Universitaire Pers Rotterdam, 1989), pp. 177–202. Quotations of Alberti and Vasari are in Elizabeth Gilmore Holt (ed.), *A Documentary History of Art* (New York: Doubleday, 1957), vol. I, pp. 212 and 230 (Alberti), vol. II, p. 26 (Vasari).

On Michelangelo and Copernicus, see Valerie Shrimplin-Evangelidis, "Sun-Symbolism and Cosmology in Michelangelo's Last Judgment," *Sixteenth Century Journal,* 21, No. 4 (1990), pp. 607–644, and Valerie Shrimplin, "Michelangelo and Copernicus: A Note on the Sistine 'Last Judgment,'" *Journal for the History of Astronomy,* 31 (2000), pp. 156–160, which she expanded in her book, *Sun Symbolism and Cosmology in Michelangelo's "Last Judgment"* (Missouri: Truman State University Press, 2000). See Edward Rosen, *Copernicus and the Scientific Revolution* (Malabar, FL: Robert E. Krieger Publishing Company, 1984), pp. 66–69 on the myth the influence of neoplatonism.

4
The Silence of Scientists: Venus's Brightness, Earth's Precession, and the Nebula in Orion

The deliberate suppressing of data contrary to one's theory is considered fraud in science today. There is, nevertheless, a gray region between the legitimate assessment of the validity of some data and the outright dishonest tossing out of bona fide information contrary to one's belief system (see section 1.3). Here are three stories of the suppression, or near suppression, of data contrary to the theories held strongly by Copernicus and Galileo.

4.1. Ptolemy on Venus

Venus, when visible, is the brightest object in the sky—after the sun and the moon. This is the reason why it probably was the first planet studied systematically by prehistoric astronomers.

Venus follows a repeatable and therefore predictable course of motion through the sky from our fixed point of view. Sometimes it rises in the east before the sun, whence it is called the morning star, lasting about 263 days; it is then lost in the glare of the sun for about 50 day, after which it reappears in the west and sets after the sun as the evening star, for another 263 days; after which it is lost in the sun's light again, this time for about 8 days, until it reappears in the east as the morning star, to begin this 584-day cycle (263 + 50 + 263 + 8) once more (note the 584-day synodic period in Table 3.1). For reasons discussed forthwith, Venus holds a special place in the history of our knowledge of the solar system, from Ptolemy through Copernicus to Galileo.

In his *Almagest,* the ancient compendium of observational astronomy, Ptolemy continued the earlier Babylonian tradition of mathematically modeling the motions of the sun, moon, and planets in order to predict their forthcoming positions. In ancient Babylon, astronomy sustained astrology, hence the compulsion to predict celestial events, but without a desire it seems to fathom the physical nature of the heavens. Ptolemy, too, had his hand in astrology, since it flourished in the Roman Empire during his lifetime, but he kept the topic apart from astronomy. There is no astrology nor mention of it in the *Almagest*; instead, he wrote another, separate, book on the topic. The *Almagest* was primarily a treatise on a mathematical model of heavenly motions, with minimal discussion of any physical structure; instead, he speculated on this structure in a shorter work, the *Planetary Hypotheses* (see section 2.3). In it his aim was to sketch what a physical model of the universe might (hypothetically) look like by assuming the planets being attached to physical spheres; he made the further assumption that all the spheres from the moon to the stars were tightly nested together, with virtually no spaces between them. By drawing the planetary circles to scale, it is possible to obtain astronomical distances.

Drawing a scale model of the major motions of Venus, ignoring the minor motions, is straightforward (see Fig. 2.3). Observationally the planet stays near the sun in its yearly motion through the zodiac, moving a maximum of about 46°. Interpreting this in Ptolemy's geocentric model results in a scale model, where the period of the deferent is a solar year (since Venus stays with the sun in its motion through the ecliptic), and the epicycle has a period of 225 days. Interestingly, the model accounts for asymmetry of the 584-day cycle—263 + 50 + 263 + 8 days— revealing why Venus disappears (50 and 8 days) when it is in the line of sight of the sun. (I leave it to the reader to see this by looking at Fig. 2.3.) Thus it is sufficient to predict the major motions of Venus. Ptolemy, as noted, used this model in the *Planetary Hypotheses* to measure astronomical distances.

Does this mean that the model is reality? That is, did Ptolemy think this scale model was a model of the actual structure of the heavens? Venus was the key planet for answering this question. Over the course of its cycle toward and away from the sun, its relative distance from Earth changes considerably, more so than any other planet. From the geometry of Figure 2.3, and considering the maximum and minimum distances for Venus from Earth, we obtain a range of relative distances of about 1 to 7 from perigee to apogee (closest to furthest from Earth, respectively). Of course, Venus is not visible when at the maximum and minimum distances (being in the glare of the sun), so its actual visible change of distance is less than this; the relative change is about 1 to 6, rounding this conservatively to a whole number. But this leads to a problem. If we conceive of Venus visually as a disk, then by the geometry of similar triangles the relative sizes of the subtended angles of the planet are proportional to their relative distances, namely 1 to 6. From this it follows that the relative differences in areas of Venus, between its closest visible distance to Earth and its farthest, should be at least about 1 to 36 (that is, 1^2 to 6^2, since the area of a circle is proportional to the square of its radius). A prediction therefore follows from the model: the change in the relative brightness of Venus from minimum to maximum in its 584-day cycle should be, when visible, at least about 1 to 36.

How does this measure up to observation? Not at all. When visible—and away from the glare of the sun—Venus barely changes in relative brightness. When visible its brightness is essentially fixed, thus contradicting the model. Since the model does not fit the data, it does not correspond to reality. This indeed was Ptolemy's conclusion.

Why then did he still apply the model to the *Planetary Hypotheses*? I think the answer is this: When the ancient astronomers said that a model "saved the appearances," they meant that it predicted the positions of the celestial objects; they did not expect it to fit the world in all ways. After all, Ptolemy used the term *hypotheses* (namely partial theses in Greek) in the title to his small book (see section 2.3).

It is imperative that I address at once what may seem a contradiction between this hypothetical interpretation of the model, and the realistic version of the calculation of the distance among Earth, the moon, and the sun using the model (see section 2.3). My appeal is to an analogy from the history of the development of quantum theory in the early 20th century (before the maturation of quantum mechanics after about 1925); although quantum theory could not account for the full range of atomic phenomena, scientists nevertheless were convinced that in cases where it worked (regarding specific experiments), it worked because it was correct. Similarly, I believe, Ptolemy—indeed, many astronomers throughout the Middle Ages and Renaissance—used the nesting-spheres model to compute relative distances of celestial bodies in the cosmos, even as, all the while, realizing that it was flawed for at least one planet (Venus). Accordingly, over the ages, as Ptolemaic astronomy was taught and manuscripts were copied, the brightness of Venus was brought to bear as evidence for conceiving of models of the heavens as hypothetical, not real. Furthermore, it was not uncommon to put this into a theological context. The argument went thusly: we are forever limited in our knowledge of heavenly things, unless Holy Scripture tells us otherwise. In short, theology was truth; science was opinion—that is, theology was the queen of the sciences. This belief and attitude prevailed to the time of Copernicus, and after.

Stellar Brightness: Hipparchus's Scale

The relative brightness of celestial objects is based on a scale for stars invented by the astronomer Hipparchus in the 2nd century BC and popularized by Ptolemy in his *Almagest*. The scale's range is from 1 (called first magnitude, for the brightest stars visible with the naked eye) to 6 (or sixth magnitude, for the faintest seen with the naked eye). The scale is a nonlinear gradient (logarithmic, like the Richter scale in seismology); it was recalibrated in the mid-19th century such that a first magnitude star is 100 times a sixth magnitude one. Until the invention of modern instruments to measure brightness, accuracy was based on the skills of astronomers experienced in the art of perceiving relative brightness, not unlike the skills of good piano tuners today using their ears rather than a tuning instrument.

Today the scale is based on a formula involving the light's wavelength. For objects brighter than first magnitude, the scale uses negative numbers. Hence the two brightest stars, Sirius and Canopus, are -1.4 and -0.7 magnitude, respectively. Specifically relevant to this story is Venus: when visible, it is between -4.7 and -3.9, a change in brightness that is barely noticeable to most viewers.

4.2. Copernicus on Venus

The heliocentric model of the heavens, put forward by Copernicus in the mid-16th century, commenced the Scientific Revolution. This implied a rearrangement of the planets as conceived since antiquity, since (among other things) Earth was now moving among them with the sun fixed at the center (see Chapter 3).

Regarding the matter of the brightness of Venus in Copernicus's model, first we must transform the scaled version of the geocentric model to the heliocentric system. The observational data are obviously the same; only the physical or geometrical interpretation is different. The fact that Venus remains observationally near the sun is due to its motion now around the sun; hence its (geocentric) epicycle becomes its (heliocentric) orbit of the sun. And so Figures 2.3 and 3.6 do not change in scale, only the arrow for the motion of the sun in 2.3 is eliminated and replaced by one showing Earth moving around the sun (3.6); also, the sun's position is moved to the center of the 225-day circle, and, although Earth is now moving around the sun, the relative visual angles remain the same. This means, importantly, that the problem of the brightness of Venus likewise accompanies Copernicus's system; or said another way, the heliocentric model does not solve the problem of the brightness of Venus from a realist viewpoint. The heliocentric system accordingly also predicts for Venus a 1 to 36 relative change in brightness, and so this model too contradicts observation. Does this not falsify Copernicus's model, if conceived of realistically?

Contrary to some historical narratives, Copernicus *did* take his model seriously. He did not interpret his heliocentric model as just another alternative way of "saving the appearances." We know that he firmly believed Earth actually went around the sun, and, like Galileo later, he (also a Catholic) hoped to convince the Vatican to adopt what he believed to be the correct model. But he surely knew that that reality was contradicted by Venus's brightness. So how did Copernicus deal with this? Quite simply, he ignored it. Nowhere in his writings, and especially in his magnum opus, *De Revolutionibus Orbium Coelestium (On the Revolutions of the Celestial Spheres),* is there even a passing mention of this problem.

Why the total silence on Venus? Of course we do not know, since he does not mention it! But the matter certainly is ripe for speculation. The late Copernican scholar Edward Rosen spoke of it as a "prudent silence." How one wishes to interpret Rosen's judgment depends on what one reads into the word *prudent.* Galileo once commented on it, which shows that the silence of Copernicus did not render the matter inconspicuous. What he said, however, I will leave for later, since I believe it sheds light on an example of another silence from Galileo himself.

The primary reason the problem was not inconspicuous is because of something written at the opening of Copernicus's *De Revolutionibus* (1543). At the start of the book, before the preface dedicated to Pope Paul III, there is a short, unsigned commentary directed "To the Reader Concerning the Hypotheses of this Work." Most readers at the time naturally thought Copernicus wrote it, although there are clues hinting otherwise. It is written in the third person, whereas the formal preface is in the first person. Moreover, it puts forward the argument that the heliocentric model should only be conceived of hypothetically, whereas the rest of the text (as the reader soon realizes) clearly looks at the model as real; indeed, in the formal preface alone Copernicus says that he will "ascribe certain motions to the terrestrial globe" and speaks of his work as aiming "to prove the earth's motion." It was the astronomer Kepler who first argued that Copernicus did not write the introductory essay: Kepler fingered the Protestant theologian and scholar Andreas Osiander, who had corresponded with Copernicus, and most importantly assisted in proofreading the manuscript during the last phases of the printing process when Copernicus was quite ill, following a stroke. Kepler was right, as confirmed by modern scholarship.

In his short introduction Osiander contends that the motions of Earth in Copernicus's book should be viewed only hypothetically, and hence they "need not be true nor even probable." At most, the model must fit the data, or, as he says, "provide a calculus [that is, a method of calculating] consistent with the observations, [and] that alone is enough," which is another way of saying the model "saves the appearances." Osiander then supports his claim with this argument: according to the model, Venus should change in brightness much more than seen by "experience." Since it does not, the model cannot be interpreted as real. The brightness of Venus, therefore, contradicts the reality of the heliocentric system.

So Osiander lets the cat out of the bag right at the start! Before the reader has barely cracked the book, Osiander loudly announces the contrary evidence that Copernicus keeps silent about. His rationale for bringing this up was probably more theological than methodological, for he ends by saying that astronomers cannot know reality because truth only comes from divine revelation; namely, science is deferential to theology—not a surprising conclusion from a theologian.

(An aside: my metaphor of "cracking a book" is probably misplaced, since books at the time could not necessarily be cracked. They were not bound when purchased, only the pages were bought; the binding was left to the discretion of the buyer.)

Independently of the theological context of Osiander's attitude, from a modern (methodological) point of view, Osiander was right. Not that the problem of Venus fully falsified Copernicus's idea, yet it did throw some doubt upon it, in revealing an apparent anomaly inherent in the model. Copernicus could have argued that the overall structure of the model was too good to be dismantled by one (although not small!) anomaly. Nevertheless he remained silent.

An obvious question arises here: Did Copernicus see Osinander's now-infamous introductory essay? The book's publication was completed in mid-April 1543 and Copernicus died on the 24th of May. A copy from the printer at Nuremberg apparently reached him as he lie dying in Frauenburg, Prussia. Even so, we do not know how his faculties were affected by the stroke. Ultimately what he saw of Osiander's essay remains a tantalizing question.

4.3. Galileo on Venus

Eventually, in 1610 Galileo solved the problem of Venus's brightness, and in Copernicus's favor. The solution followed simply and clearly from Galileo's telescopic discovery that Venus, like the moon, goes through periodic phases. Diagrams convey this best: Figure 4.1 shows that Venus does not always "shine" with the same (constant) intensity of light because as it orbits the sun it displays phases from the viewpoint of Earth. When it is closer to us it is more of a crescent and hence it reflects less light than when farther and full; the compensating factor implies that from the vantage point of Earth, Venus always gives off about the same amount of light when visible and out of the sun's glare. It is an obvious fact today, which follows from another fact—that Venus shines by reflected light from the sun, without any internal light. But, to repeat a mantra of this book, what is obvious today was not so in the past.

In addition to falsifying Osiander's objection to the Copernican system, the discovery of the phases of Venus proved that it orbits the sun. Comparing the geometrical arrangements in Figure 4.2, it is clear that only in Copernicus's system does Venus pass through all four phases: first quarter, full, third quarter, and new. In Ptolemy's there are two new phases but no full one. This was strong evidence for the Copernican system, as Galileo doggedly noted. This alone, however, did not constitute proof of heliocentrism, since Venus was assumed to orbit the sun on Tycho Brahe's system too. Put forward in 1588 as a compromise between those of Ptolemy and Copernicus, Tycho's model kept Earth fixed at the center of the stellar sphere, with the sun revolving around it, but, in turn, he placed the planets in orbit around the sun (see Figure 5.1; more on this in section 5.2).

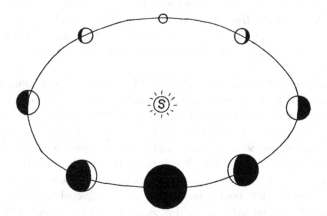

FIGURE 4.1. The brightness of Venus: Galileo. This diagram illustrates how Galileo's discovery of the phases of Venus explains why Venus (when clearly visible) shines by a constant light to the naked eye. Notice how the planet is closer at the crescent phase, yet it gives off less light than when it is full.

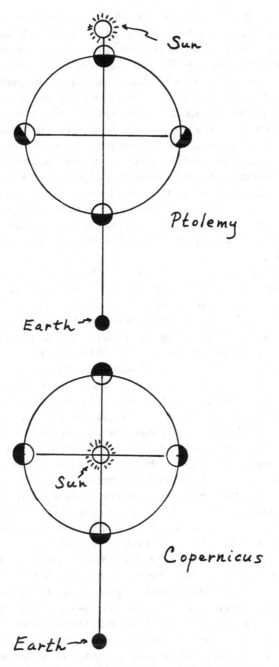

FIGURE 4.2. The phases of Venus: Ptolemy and Copernicus. A comparison of the phases of
Venus as seen on Ptolemy's and Copernicus's systems. Note that there is no "full" Venus
on Ptolemy's system. Venus goes through all four phases only if it orbits the sun.

So Figures 4.1 and 4.2 amply show how the ancient problem of the brightness of Venus was resolved—seemingly simply and clearly, as any textbook may present. But a theme of this book is the complexity of history, which does not always follow a clear-cut logical path. For example, since ancient times there was no unambiguous view on the possible self-luminosity of the planets. Although Ptolemy did not address the matter, medieval and Renaissance writers raised it, proposing four possibilities for the planets: opaque, self-luminous, trans-parent, or a combination of transparency and self-luminosity. Throughout the Middle Ages, the constant brightness of Venus was generally interpreted as evidence of its self-luminosity, ignoring in the process, however, the ensuing geometrical problem. By Galileo's time, the combination hypothesis was common. So the issue of the brightness of Venus was tied not only to its celestial placement but also to its constitution. Galileo acknowledged this in his writings on Venus. Yet the entangled matters around Galileo's work on the phases of Venus are often less about actual history and more so involving historiography—that is, debates among historians.

At the annual meeting of the (U.S.) History of Science Society in 1983, the now late and eminent science historian Richard S. Westfall presented the provocative thesis that Galileo stole from a student the idea that the phases of Venus would prove its orbit around the sun. Published in 1985 in the society's journal under the title, "Science and Patronage: Galileo and the Telescope," the paper went on to win the journal's award as the outstanding article of the year. The paper is a mine of information on the social conditions of scientists at the time. Working from the texts of Galileo's writings and correspondence, Westfall paints a fascinating picture of Galileo's scientific pursuits within the context of securing the favors of patrons—all as a means of trying to leave his low-paying job as professor of mathematics at the University of Padua. The ploy worked; he named the four moons of Jupiter that he discovered with his telescope in January 1610 after the four Medici brothers, for which he obtained a prestigious position at the Medici court in Florence, to which he moved in September 1610 (see section 7.1).

Westfall's accusation is grounded on a letter Galileo received from a former student, Benedetto Castelli, dated December 5, 1610, in which the student alerts Galileo to the possibility of Venus exhibiting phases, which, he points out, would settle the issue of Ptolemy versus Copernicus. Westfall assumes that Galileo had neither thought of this possibility nor had he looked at Venus through his telescope before receiving this letter. From Westfall's viewpoint, Galileo at the time was not pursuing a consistent observational program of the sky; instead, he used the telescope primarily as a means of procuring patrons—note the juxtapositioning of the main title (on patronage) and subtitle (on the telescope) of Westfall's paper. He writes, "Galileo seems to have used his telescope to further his advancement rather than Copernicanism." Indeed, Galileo did not reply to his student until December 30, when he said that he had been observing Venus in detail for the past three months and that he had, indeed, discovered the phases. But Westfall contends that Castelli's suggestion actually inspired Galileo to look at Venus for the first time. Westfall reinforces his case by pointing to the fact that, in the meantime, on

December 11, Galileo sent an anagram to the astronomer Kepler, which contained the (predicted) discovery of the phases of Venus. In the context of his new job, there was considerable pressure on Galileo to make more discoveries for the court; he did not want someone else to announce the discovery before him, but he also did not want to be wrong—hence the cipher. (I should point out that the use of ciphers was common practice at the time, rather like prepublication notices today, but with an air of mystery.) The encrypted anagram read, *Haec immatura a me iam frustra leguntur o y*, which means, "These, premature from me, are at present deceptively gathered together." The decoded anagram read: *Cynthiae figuras aemulatur mater amorum*, which means, "The mother of love [Venus] imitates the shapes of Cynthia [the moon]."

Astronomer-historian Owen Gingerich and others quickly challenged Westfall's thesis. They reconstructed how Venus actually appeared from the summer through December 1610, by slightly blurring the images to conform to what Galileo may have seen in his crude telescopes. Paolo Palmieri recently made a more meticulous visual reconstruction. What his images show is that there was nothing particularly interesting about Venus in the summer and into early fall, but beginning in October its shape began to change, appearing slightly flattened on one side. It was not until near the end of December that visual confirmation of the phases would have been attainable, as Venus clearly approached a crescent phase, just when Galileo replied to his student.

In contrast to Westfall's thesis, the more plausible scenario is this: from November 1609, when Galileo first turned his telescope to the moon, he consistently observed the night sky. In January he studied Jupiter, as it was then visible, and he discovered the four moons. Mars was later visible too, but nothing special to behold. Saturn was not visible until July, when we know Galileo observed its "strange countenance," with lumps in either side. (Over 40 years later, the Dutch astronomer-mathematician Christiaan Huygens would resolve the rings of Saturn.) In the summer, as Venus became visible in the evening sky, Galileo turned his telescope to it when in Padua; after moving to his new job in Florence, he resumed his observations, just about in time for the transformation of the planet. Upon seeing the flattening of the side he became more diligent in his viewing. In the letter to Castelli, Galileo says he has been observing Venus in detail for the last 3 months; this is consistent with his having casually observed the planet since the summer but only systematically since October. Hence there is no reason to accuse Galileo of dishonesty.

On January 1, 1611, Galileo revealed to Kepler the meaning of the cipher. Importantly, and correctly, he declared that this discovery solves two "great questions" at once: the issue of Venus's orbit of the sun and the nature of the planet, which unquestionably is opaque—implying that the other planets are opaque, too. He was particularly pleased to resolve the latter, which "up until now was unclear to the greatest minds of the world" ("*sin qui dubbie tra i maggiori ingeni del mondo*").

As a coda (or is it the clincher?) to this story, I add this account of Gingerich's face-to-face meeting with Westfall (which appeared in Gingerich's obituary of

"Sam," as Westfall was know to friends). Gingerich writes: "I pointed out [to Westfall] that Galileo would surely have looked at Venus as soon as it was visible in the evening sky, since it is the third brightest astronomical object after the sun and moon. 'Really?' Sam exclaimed. 'But I am not astronomer!'"

If there is a moral to this tale, it is that historians of astronomy should not rely exclusively on written texts. Or, said another way, they should get outside more often and look at the heavens.

4.4. Galileo, Sunspots, and Precession

Having exonerated Galileo from Westfall's accusations, I now turn the tables on Galileo and accuse him of being less than honest on a different matter, in the course of which I return to the topic of the title of this chapter.

In his *Dialogue on the Two Chief World Systems* (1632), the book that got him into deep trouble with his church, Galileo put forward his best arguments for the Copernican system. (The book's format is a "dialogue" over four days among three interlocutors, so Galileo is ostensibly giving equal time to both viewpoints, helio- and geocentrism.) Near the end of the book (on the fourth day) he sums up the "strong evidence in favor of the Copernican system," by presenting three "very convincing" phenomena: the explanation of retrograde motion, the theory of the tides, and the motion of sunspots. The first is the most well known and is found in most textbooks today. From a heliocentric point of view, the planet's retrograde motion is more simply explained by the motion of Earth bypassing the planet; thus the planet (geocentrically) only appears to move backward (see Fig. 3.5). A comparison of the two viewpoints displays what became known as the relativity of motion. The second, involving the tides, makes the case that the back and forth motions of the tides are caused by the simultaneous rotation of Earth on its axis and its likewise revolution around the sun; moreover, if true, any mysterious (action-at-a-distance) powers or forces between Earth and the moon purported as causing the tides, such as was speculated by astrologers and other mystics, were purged and therefore Galileo believed he had secured gravity as being only a local power near Earth. Of course, Newton's tidal theory ultimately replaced Galileo's, which because of its errors is usually dismissed in historical writing. The third—and the focus of attention here—is the sunspots argument, seldom found in textbooks and probably the least studied "proof." Nevertheless, there have been commentaries and scholarly probes of Galileo's argument, but they are fraught with much confusion and debate. Part of the reason is the admittedly (by Galileo himself) slightly confusing and obscure presentation of it in the *Dialogue*. There is evidence that this is at least partially because he inserted the sunspots proof as the last piece in his manuscript. Whereas he had been pondering the proof from the tides for years, it seems he came upon the sunspots proof rather suddenly when finishing the manuscript and it was inserted during the final stages of publication.

Galileo was not the first to see sunspots; there are reports of possible markings on the sun going back at least to the Middle Ages. Nor was he the first to study them and their motions with a telescope. The heliocentric interpretation, however, was uniquely his. He first mentions sunspots in a letter to his artist friend, Lodovico Cigoli, October 1, 1611 (see section 7.1), in which he implies that he has been observing them for some time; historians think he probably first saw them early in 1611. His first discussion of sunspots centers on their nature and the question of their apparent contradiction with the idea of the incorruptibility of the heavens, as held by Aristotelians. This led to one of the major scientific disputes in Galileo's life, for the idea of the perfection of the heavens was firmly entrenched among Aristotelians, and the sun was an especially powerful symbol of this flawlessness. For example, one alternative interpretation was that the "spots" were really small planets in orbit between Mercury and the sun. In defense, Galileo drew on his knowledge of perspective and art, pointing to the fact that the sunspots change their shape as they move across the face of the sun, such that they become narrow near the edges; he interpreted this as a foreshortening of the spots, implying that they are *on* the sun's surface.

Following the motion of the sunspots he concluded that the sun is rotating with an approximate 1-month period. His systematic study further revealed a changing pattern of motion across the sun's surface over the course of a year. It was this pattern that he found was most naturally explained on the heliocentric model. The proof is quite strikingly seen with the help of diagrams. Over a year there are four patterns of sunspots from left to right across the sun, as shown in Figure 4.3: a upward path, an arc, a downward path, and another (opposite) arc. Assuming the sun's axis is tilted to Earth's plane around the sun, as in Figure 4.4, then with a little visualization from four different seasons (A, B, C, D), it is possible to "see" the four monthly patterns of Figure 4.3. It is no wonder that he declares this heliocentric interpretation of the data to be "a more solid and convincing theory of the sun and earth than has ever been offered by anybody." Indeed it is all the more remarkable if indeed Galileo came upon this proof not long before completing the manuscript. As it stands, the proof seems ironclad.

But, of course, given the format of the dialogue, the geocentric viewpoint demanded equal time. And so voice is given to an explanation of the observed pattern from a fixed Earth. The result is complex and difficult to conceive. It involves four different motions for the sun; particularly problematic, according to Galileo, is that these motions are both clockwise and counterclockwise—and thus "incongruous with each other."

Here are the details. There has been some confusion, as well as downright errors committed by scholars, about the motion of sunspots as seen from a *fixed*

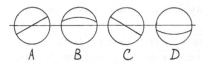

FIGURE 4.3. Sunspots: yearly patterns. There are four distinct patterns (A, B, C, D) of sunspots across the face of the sun over approximately equal intervals during the course of a year.

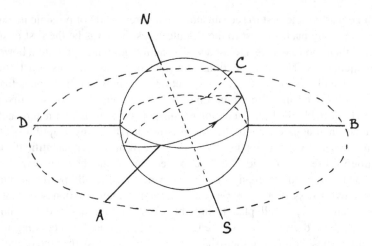

FIGURE 4.4. Sunspots: heliocentric viewpoint. This diagram provides an explanation of the patterns of sunspots (see Fig. 4.3) on the heliocentric model. Notice how the four positions of Earth around the sun (A, B, C, D) correspond to the four patterns.

Earth. Indeed it takes quite a bit of mental gymnastics to conceptualize Galileo's argument against the geocentric explanation of the motion of sunspots. A diagram (Fig. 4.5) helps. First, the sun (S) must make a daily circuit around Earth; the celestial equator revolving clockwise around the North Celestial Pole (N) produces this motion. Second, the sun makes an annual circuit counterclockwise along the ecliptic, which is tilted 23½° to the celestial equator. Compounding these two motions produces path P (parallel to the celestial equator) on a given day; accordingly, over the course of a year, path P moves north and south, between the solstices, being coincident with the celestial equator at the equinoxes.

Third (and this is perhaps the trickiest one to conceive), the sun must also perform an annual clockwise conical motion around an axis perpendicular to the plane of the ecliptic; this motion is required to explain the fact that the sun remains tilted at the same angle throughout the year. And, fourth, the sun monthly rotates counterclockwise on its own axis. Hence the sun makes four motions— two clockwise and two counterclockwise. Is it any wonder that Galileo declares this model to be too cumbersome to be true?

In contrast, the heliocentric model involves only Earth rotating (daily) on its axis and, importantly, in the *same* direction as it revolves (annually) around the sun (counterclockwise): in Galileo's words, "two simple noncontradictory motions [are] assigned to the earth." The differences between Ptolemy and Copernicus thus involve contrasts: complexity versus simplicity, and incongruity versus congruity, respectively.

At this stage of the argument Copernicus seems to win, hands down. Nevertheless, when all necessary motions are considered, in order to preserve

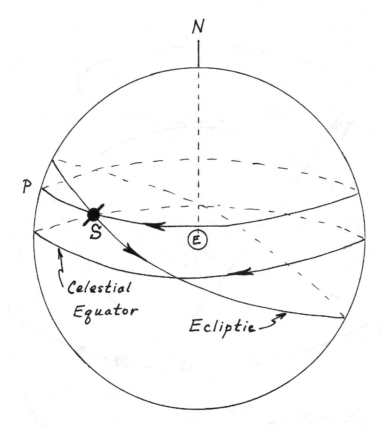

FIGURE 4.5. Sunspots: geocentric viewpoint. This geometrical construction (placing the celestial equator horizontal in the diagram) is helpful for seeing how the motion of sunspots may be explained on the geocentric model. Four motions of the sun (S), two clockwise and two counterclockwise, are required to produce the phenomena as seen from a geocentric viewpoint. They are a daily clockwise circuit along path P around Earth, an annual counterclockwise circuit along the ecliptic, an annual clockwise conical motion, and a monthly counterclockwise rotation.

the aesthetic quality of the Copernican system (with its simplicity/congruity), Galileo must perform a slight of hand. To explain, we must return to ancient astronomy.

About 1800 years before the *Dialogue,* Hipparchus discovered that the equinoxes (spring and fall) come earlier (preceding) every year by a very, very small amount; hence the phenomenon was dubbed the "precession of the equinoxes" (see section 3.5). On the ancient geocentric model, this observation required the addition of another motion (and hence another sphere) for the heavens; in this case a conical motion exercised by the celestial sphere was sufficient. It may take a little mental gymnastics to see this too. In Figure 4.6 I have drawn the ecliptic

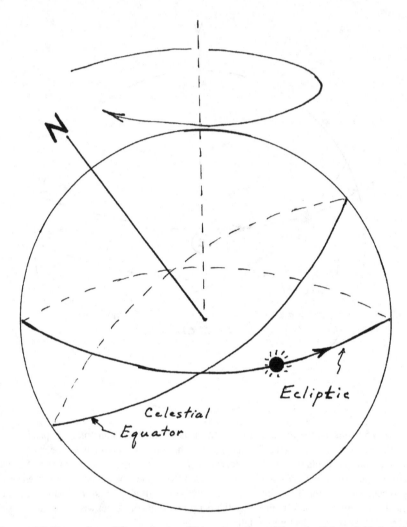

FIGURE 4.6. Precession of the equinoxes. This orientation of the celestial model is helpful for explaining the precession of the equinoxes from a geocentric viewpoint. The ecliptic is fixed and the celestial sphere makes a conical motion as displayed by the looping arrow. Accordingly, the North Celestial Pole makes a loop over the 26,000-year period (see Fig. 4.7).

horizontal to the page, which I hope makes this motion easier to see. First mentally fix the ecliptic; now notice that if the sun is to arrive at the equinox a bit earlier each year, then the celestial sphere must perform a conical motion as indicated. Since this motion is continuous, the conical motion is periodic, taking 26,000 years for a cycle. (Hipparchus apparently rounded off his data to 1°/century, resulting in a 36,000-year period that was generally used over the ages. Galileo uses the 36,000-year period in his *Dialogue,* although Copernicus had computed a value of 25,816 years in *De Revolutionibus*.) The value is extremely large because the yearly "preceding" equinox is very small. True, the precession is small, but not irrelevant. In the ancient (geocentric) cosmos, therefore, there were usually two spheres beyond Saturn: one performing this conical motion and another performing the requisite diurnal (daily) motion, since *all* visual phenomena must be accounted for.

Switching to the heliocentric model, where the celestial sphere is fixed, a conical motion, therefore, must be assigned to Earth to account for precession (just as a daily motion is assigned to Earth to account for the diurnal motion of the heavens). Specifically, in addition to its daily rotation on its axis and an annual revolution around the sun, Earth also performs a conical motion about an axis perpendicular to the plane of its orbit (similar to that in Fig. 3.7, except with a 26,000- or 36,000-year period, rather than 1 year). Indeed, because of this, all such conical motion is now called precession, such as a rotating top; see essay on Newton and precession (section 11.1).

Now here's the kicker: whereas the two motions of Earth that account for the day and the year on the Copernican model are both in the same (counterclockwise) direction, and hence accommodate Galileo's aesthetic criterion of congruity, the conical motion necessary to explain the precession of the equinoxes, on the contrary, is clockwise. So Earth must perform three motions (two counterclockwise, one clockwise) on the Copernican model. This means that the very lack of harmony that led to the falsification of the Ptolemaic model for the motion of sunspots—recall Galileo complaining of the "incongruity" among the motions—now appears in the Copernican model when precession is added. A key feature of the aesthetics of Copernicus's model has evaporated. Here, then, is an important problem that Galileo must confront. How does he deal with the "incongruous" nature of these motions?

Like Copernicus in his stratagem on the problem of the brightness of Venus, Galileo plays the silent game. Nowhere in the *Dialogue* does he explicitly confront the reader with the fact that a conical motion of Earth is required to explain the precession of the equinoxes. He does, though, mention the phenomenon of precession, but almost exclusively within the context of the Ptolemaic model. Moreover, it is brought up in order to find fault with geocentrism, for the motions (precessional and diurnal) of the last two spheres break the harmony of the model. How ironic. Or, how cynical?

Let's look more closely as what Galileo says about precession. It is mentioned only four times in the *Dialogue*. It first appears within a discussion of the ordering the planets and stars, and he notes that beyond Saturn in the Ptolemaic model "many thousands of years" (i.e., the 36,000 years) are required for the period of the stellar sphere that accounts for precession; beyond this is the last sphere accounting for the daily motion, thus destroying any hierarchical and sequential (temporal) order. That is all he says; the fact that this must translate into a motion for Earth on the Copernican model is not disclosed.

The second appearance of precession is within an attempt to calculate the distance of the stars based on the hierarchical ordering of the celestial bodies with increasing periods of revolution from the center. The context of the discussion is the anti-Copernican argument that if Earth really moves, then stellar parallax should be observed (see section 2.3). Galileo's defense is to provide evidence for an enormous distance to the stars. The calculation rests on what appears as a rather strange fusion (or is it a confusion?) of both models.

On Copernicus's model, setting Earth–sun distance to unity (1 AU) makes Saturn, for example, 9 AU from the sun (see section 3.3). Using the Ptolemaic system, the relative distances are obtained using the nesting-spheres model (see section 2.2), and Ptolemy arrived at a range of values (for example, Saturn was 12 to 17 AU, with a central Earth). Now here's where Galileo's fusion/confusion arises. Noting the hierarchical correlation between planetary distances and periods (see section 3.2 and Table 3.2), he adds the stellar sphere, using the 36,000-year precessional motion as its period. Even though Copernicus did not abandon the concept of a stellar sphere, by using the 36,000-year precessional motion for the sphere Galileo is, in fact, incorporating the geocentric viewpoint, since Earth's conical motion accounts for the precession of the stars on Copernicus's model and hence the stellar sphere is fixed. But Galileo proceeds by using Copernicus's data for the following calculation, which thus mixes the two models. He derives the distance to the stars this way: since the period of Saturn is 30 years and the period of the stellar sphere as 36,000 years, then the equation $9/30 = x/36,000$, yields $x = 10,800$ AU for the distance of the stars. He makes similar calculations using Jupiter and Mars and arrives at two more large numbers, 15,000 AU and 27,000 AU, respectively (he errs here; the latter should be even larger, about 28,700 AU). These give him what we might call ballpark figures (or a range thereof) for the stellar distance, large enough, at once, to refute the anti-Copernican argument and supposedly support Copernicus.

Can this mixing of the geo- and heliocentric models be justified? If Galileo used Ptolemy's numbers and model (of nesting spheres), he would necessarily set the stars just beyond the sphere of Saturn and of course not support Copernicus. Only on Copernicus's model is there a gap between Saturn and the stars. Nevertheless the 36,000-year period for the stars may only be applied to the geocentric system, since Earth performs this precessional motion on the heliocentric model, but, as noted, this Galileo conveniently ignores. Thus the fusion of the two models renders the entire argument and calculation spurious, at best. If, as I suspect, he was fully aware of this, then there was no confusion, nor even a delusion; we are, instead, observing a classic case of sophistry.

Lastly, and not surprisingly, he again points to the lack of an ordered hierarchy between the last two spheres (since the one after Saturn, having a 36,000-year period, is followed by the last with only a daily period, thus destroying the hierarchy); these he calls "monstrosities." In the end, and for a second time, a precessional motion is noted, but without acknowledging this motion for Earth.

Incidentally, the alert reader will note that Galileo's calculation is wrong from another (although probably anachronistic) viewpoint: he is assuming a direct proportion between the periods and the distances of the planets. Kepler's third law, however, states that the square of the period is proportional to the cube of the distance. This he published in 1619, but it was not yet commonly known when Galileo published his *Dialogue* (1632), since the law was buried in his treatise on *The Harmony of the World* that many, and Galileo in particular, considered essentially a foray into mysticism (see section 6.4).

One of the visual manifestations of precession (that is, observationally, and hence independent of the system used) is that the North and South Celestial Poles are not fixed in the heavens, but over the 26,000-year cycle (using today's value) they each make circles in the sky. For the Northern Hemisphere, due to the conical motion in Figure 4.6, from the fixed Earth the North Celestial Pole rotates as in Figure 4.7. In other words, that there is now a pole star (called appropriately, for a time, Polaris) near the North Celestial Pole is merely a chance occurrence, just as there is no pole star at present in the Southern Hemisphere. (In about 12,000 years

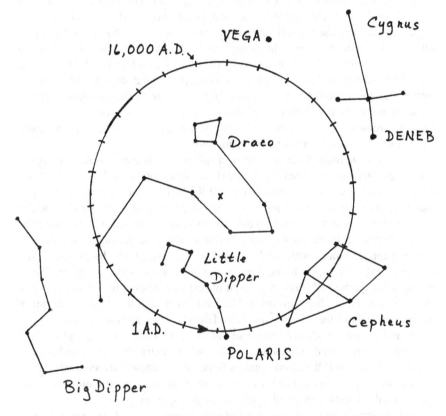

FIGURE 4.7. Motion of the celestial pole. This diagram charts the counterclockwise motion of the North Celestial Pole through the stars over its 26,000-year period. Although Polaris is now the North Star, it was not and will not always be so. For example, in about 12,000 years Vega will be the closest star to the Pole. The closest it gets to Polaris will be in 2102.

Vega will be the pole star in the Northern Hemisphere, or at least will be the nearest star to the pole.) This motion of the poles is mentioned in the *Dialogue* in the third reference, where the requisite motion of Earth is acclaimed (for the first time!) in this phrase: "It is true that such points in the heavens [i.e., the Celestial Poles] are changed when the transposition of the earth is carried out in such a way that its axis points to other parts of the immovable celestial sphere." That's it, a phrase, not even a complete sentence, and no more is said. Galileo fails to inform the reader that "the transposition of the earth" specifically entails a conical motion within the Copernican model.

The fourth and final reference to precession is the most direct. It appears in the context of a discussion of the tilt of Earth in the Copernican model and the fact that the additional ("third") motion Copernicus postulated for Earth is now unnecessary (see section 3.5). Let me explain this before specifically looking at this last reference to precessional motion.

At this place in the *Dialogue* Galileo presents a wonderful argument for eliminating the third motion of Copernicus (see Fig. 3.7). The background to it is twofold. In the late 16th century, the Danish astronomer Tycho Brahe made a strong case against the real existence of the celestial spheres based on the apparent path of comets moving right through them; this left physical astronomy (on any model) with a problem: What keeps the planets in orbit? In addition, in the early 17th century Galileo arrived at the concept of what later would be called inertia to explain motion (see section 8.2). Essentially Galileo realized that any object in motion stays in motion or stays at rest unless changed by an external force or pressure (say, a medium slowing it down if it is moving, or by some direct impact moving it or changing its motion).

Now in the *Dialogue* Galileo submits a splendid little experiment that anyone can do and that has direct bearing on the third motion problem. Here it is: all you need is a bowl of water and a ball; it helps if the ball has some marks on it so you may keep track of its orientation. Carefully place the ball in the bowl of water, and, keeping it near the center (avoid having it touch the side), now rotate the bowl with respect to the room. You will find (perhaps surprisingly) that the ball does not move (rotate) with the bowl and water; instead, the ball's orientation remains fixed with respect to the room as the bowl and water are rotating. This is inertia: namely, the ball stays in its state of rest with respect to the room despite the motion of the medium around it. Galileo then applies this to the Copernican model, but without the spheres, this way. Earth has two motions around the sun (daily and annual); it also remains fixed at the $23\frac{1}{2}°$ to the perpendicular to the plane of its orbit. And due to inertia, Earth needs no extra (third) motion because, like the ball in bowl, it indeed remains fixed at this angle independently of the other motions (see Fig. 3.7, without the conical motion). Inertia eliminates the need for a third motion in the absence of celestial spheres.

From today's viewpoint this is an obvious application of an experiment to solve a potential problem with a model. In the 17th-century context, however, Galileo's application was quite radical. Recall that in the Aristotelian framework there was an essential distinction between the two cosmic worlds: terrestrial and celestial. Applying something from one realm to the other was thus meaningless. So arguing that the behavior of a ball in water on Earth had relevance to the motion of the planets in the heavens was most probably seen as rather comical— and, of course, dead wrong. But Galileo's Copernican framework entailed the breakdown of that division, since the terrestrial "world" (Earth and the moon) was now in orbit around the sun. The cosmos was one—a holistic, homogeneous view, seen, for example, in the moon's surface being more like Earth than ever thought (see section 7.1).

This finally brings me back to the precession of Earth and the fourth reference. It appears in one sentence, as follows: "This [tilt of Earth] remains perpetually the same, except for some small variation in many thousands of years which is not significant in the present connection." The "connection," namely the context of the discussion, is Galileo's realization that Earth remains tilted by inertia, and hence Copernicus's extra motion is unnecessary; indeed, it is to this topic that the discussion returns, focusing on the "immutable" tilt of Earth, without, accordingly, pursuing the matter of the "small variation" in the tilt. Being such a small motion compared to the diurnal and annual motions, Galileo thus dismisses Earth's precession as "not significant." And that is *all* he says on the precession of Earth.

There is another way of looking at this last reference. Galileo's removal of the third motion of Copernicus may have been a factor in his silence on Earth's precessional motion on the Copernican model. As seen, Copernicus used the third motion to account for the precession of Earth, by giving it a slight slippage. Without the third motion, Galileo had no feasible way to add the precessional motion; there was nothing to give a slippage to. Consequently, he had to add another entirely new motion—really a different (although very small) third motion—to account for the precession of the equinoxes. Thus the relative silence?

To summarize, of the four references in the *Dialogue* to precessional motion, only the last two mention Earth's motion: an allusion in a passing phrase and a dismissal in a single sentence. That's Galileo total output on the conical motion of Earth on the Copernican system, a topic to which Copernicus devoted a significant part of his *De Revolutionibus*.

This is not quite the silence of Copernicus on the brightness of Venus, but nearly so. Interestingly, Galileo comments on Copernicus's silence about Venus in the following sentence, which, if I am permitted some psychological speculation, may be read as justification for his own relative silence on precession: "I believe this [silence] was because he was unable to rescue to his own satisfaction an appearance so contradictory to his view; yet being persuaded by so many other reasons, he maintained that view and held it to be true." Put another way, Galileo and Copernicus looked toward the "big picture" and thus avoided or ignored small anomalies that contradicted their greater visions. A similar justification appears earlier in the *Dialogue* when Galileo discusses Copernicus's silence on the problem of how birds can keep up with a moving Earth. I think we may find the same subconscious significance in this declaration: "Perhaps Copernicus himself was unable to find a solution which entirely satisfied him, and for that reason he remained silent on it." How intriguing and possibly insightful it is to observe Galileo explicitly commenting on the silence of Copernicus, while we catch him in the act himself.

Galileo's own dismissal of the precession of Earth is only exceeded by the silence of the handful of scholars who have studied his argument on sunspots. Several years ago, when pursing research on this topic and discovering Galileo's dismissal, I searched in vain for someone else to have noticed Galileo's lack of reference to this motion of Earth. Surprisingly, even the most recent article on the topic by Paul Mueller, seemingly the most detailed to date, perpetuates the silence by ignoring precession.

Mueller's article was published after my article disclosing Galileo's dismissal appeared; specifically, Mueller saw my article while his was in press, and he curtly dismissed my argument in a footnote.

His rejection is based on his belief (and he admits it is a belief) that Galileo made a clear and sharp distinction between two criteria for assessing models— simplicity and congruity. Moreover, according to Mueller, Galileo himself only believed in the simplicity criterion; congruity was the criterion of the Aristotelians. So in the *Dialogue* Galileo cleverly set a trap for the Aristotelians by showing that the Ptolemaic model was incongruous and the Copernican was simple; Mueller writes, "[If the Aristotelians] opt for the Ptolemaic explanation of sunspot motions, they will have to admit that an astronomical system with incongruous motions can be physically plausible. If on the other hand they embrace the Copernican explanation of sunspot motion, they will have to admit that a mobile Earth is physically plausible." Mueller's thesis is a most seductive one, to be sure. He then turns it against me, since I show that precession of Earth involves incongruous motions; but, according to Mueller, this has no bearing against Galileo, since he does not really believe in this (Aristotelian) criterion. It is as if Galileo has special dispensation: he may wield the verbal weapons of simplify and congruity in the mouths of the three discussants in the *Dialogue* at will, but I may not turn one (incongruity) on him; he is impervious to its arrows. Contrary to Mueller, I find it exceedingly difficult to be this generous to Galileo, however much I would like to be.

4.5. Galileo and Nebulae

As a coda to this chapter's topic, here is yet another case of Galileo's silence. It appears in his *Sidereus Nuncius,* in his discussion of the nebulae sighted through his telescope. Whether in ancient times or now, about half a dozen or so nebulae are visible with the naked eye. Because they appear as blurred or fuzzy stars, ancient astronomers called them nebulous stars (see section 12.1). Ptolemy lists six and Copernicus five in their catalogues of the stars. Galileo describes his telescopic viewing of them right after his explication of the Milky Way. Ever since ancient times the Milky Way was thought to be made of vapors suspend below the sphere of the moon because of its continuum-like appearance and its changing countenance over the year. But Galileo's telescope resolved its discrete nature. As he writes, "The [Milky Way] Galaxy is nothing else than a congeries of innumerable stars distributed in clusters." Having found the apparent continuum of the Milky Way to be really discrete stars packed closely together, he goes on to reveal what he saw of the nebulae. "Moreover—and what is even more remarkable—the stars that have been called 'nebulous' by every single astronomer up to this day are swarms of small stars placed exceedingly closely together." As is the Milky Way, so are the nebulae. Specifically he describes, and provides diagrams of, two nebulae: that in Orion's head, and Praesepe (today the Beehive cluster, M44) in Cancer, both mentioned by Ptolemy and Copernicus.

This commentary on the Milky Way and the nebulous stars follows his discussion of the innumerable stars revealed by the telescope beyond naked eye observation. As an example, he discusses, with a diagram, the many stars in the Pleiades, beyond the seven visible with the naked eye (the so-called Seven Sisters). Relevant here is his viewing of the constellation Orion, which he says he first intended to depict completely, "but overwhelmed by the enormous multitude of stars and a lack of time, I put off this assault until another occasion [which never transpired]. For there are more than five hundred new stars around the old ones." In the end he confines himself to depicting the numerous stars around the three visible stars in Orion's belt and the six in the sword: "I have added eighty others seen recently, and I have retained their separations as accurately as possible." The point of all this, of course, is the appreciation of the multifold nature of the stars beyond the few thousand we see with the naked eye. Nevertheless, Galileo's actual diagram of Orion's sword, to a knowledgeable viewer, poses an obvious question: Where is the nebula? For one of the conspicuous nebulae in the winter sky is that found in Orion's sword. But it is not in Galileo's diagram nor is it mentioned in the text. Why not? Surely, as he carefully counted the stars in Orion's belt and sword, he saw this nebula through his telescope. So why is he ignoring it?

Before presenting my answer, I should point out that another quite interesting explanation for this has been proposed: namely, that the nebula was not really there in the early 17th century! Instead it was a star that later exploded. This is logically a plausible answer; however, Owen Gingerich, in a well-research article, has convinced me that there is no evidence to support this thesis.

If, as seems to be true, the nebula was indeed refracted through Galileo's telescope in 1610, why then the silence? We know today that this nebula is really what we still call a nebula, that is, a region of hot gases, not a star cluster. So when Galileo set his telescopic sights on it, expecting no doubt to see a star cluster (as were the Milky Way and the other two nebulae mentioned above), he was either surprised or disappointed (or both) that it remained a visual continuum. He may have believed that his telescope was just not powerful enough to resolve the individual stars or he may have thought that this nebula really was a continuum. In either case—or perhaps he had some other hypothesis, it does not matter—he does not mention it. His silence, therefore, seems to be based on his belief that this nebula is either temporarily or potentially a counterexample to his newly concluded conjecture on the discrete nature of the nebulous stars, and he is not predisposed to question it so soon. Why disclose a counterexample that may be explained away in due time? In this instance, at least, I am willing to concede that Galileo's silence may be labeled as being "prudent."

Finally, another intriguing, albeit tentative, hypothesis has been put forward by Thomas R. Williams of the Galileo Project at Rice University in Houston. Attempting to reproduce Galileo's results using a similar telescope, he concluded that perhaps Galileo did not even see the nebula because of the small field of view of his telescope. Many of Williams's students who used a "replicate" telescope were unable to see the nebula, although he concedes that light pollution (not a problem for Galileo) may be factor today.

Galileo and Holy Scripture

Galileo was fond of quoting the aphorism (not of his own coining) freely translated as: "Religion tells you how to go to heaven; science tells you how the heavens go." I believe Galileo embraced this viewpoint that science and religion have dominion over different realms and hence should not inherently be in conflict. It could thus be used as an excuse to stay out of any dispute between religion and science. Nevertheless, we still find Galileo wading into the fray of debating the apparent clash between passages of Scripture and the Copernican cosmology (some scholars say he was dragged into it).

A well-known example involves the biblical passage in the Book of Joshua, where Joshua commands the sun to stand still in order to prolong the day; theologians obviously interpreted this as embodying the truth of the Earth-centered model with the sun moving around it, not vice versa. Galileo's response is so clever that when I first read it, I laughed out loud.

Galileo begins by making the case that we must take the words of Holy Scripture as absolutely true since the word of God cannot be a lie. Therefore, Joshua did command the sun to stand still to lengthen the day. With this preamble it would seem that Galileo is painting himself into a corner, but surely he is too clever for that. There follows a lesson in astronomy, Ptolemaic astronomy. The actual and only true motion of the sun in the geocentric system is its yearly motion through the ecliptic. The daily rising and setting of the sun is caused by the motion of the stellar sphere, which is the source of all diurnal celestial motions (of the moon, planets, and so forth). Thus, to stop the day (namely, the daytime), one would have to command the starry sphere to stop, not the sun (Fig. 4.5). To further nail his point, Galileo spells out that since the yearly motion of the sun is from west to east across the sky, then to

stop the sun during the day would, in fact, stop this westward drift and consequently increase the eastward motion; in the end, stopping the sun actually
decreases the length of the day!

Galileo's point is that if we take a literal interpretation of the Bible, then it
follows that (what may be called) a literal interpretation of Ptolemaic astronomy is in order, and that is precisely what he has done.

I believe he knew that all this was really an exercise in sophistry, but he
evidently enjoyed the game. That it was sophistry is revealed, I would argue,
by Galileo's next move, arguing that Joshua's command actually supports the
Copernican model.

He presents the heliocentric model this way: drawing on his discovery of
the rotation of sunspots, he makes the case that the sun's rotation is the source
of the motions of the entire planetary system. Hence "stopping the sun"
(interpreted now as stopping its rotation on its axis) would halt all other
motions of the entire system of the planets, entailing the stopping of the day.
So Joshua's command supports Copernicus.

How are we to assess this argument? What does he mean that the sun is the
source of motion? First, Galileo is not speaking of gravity, since we know that
he wanted gravity to be locally confined to Earth, otherwise shades of occult
powers between Earth and the moon (for example, to explain the tides) might
follow, and he was adamantly against that. Kepler had put forward his model
of magnetic powers between the sun and the planets (see Fig. 11.2), but that
too smacked of occultism for Galileo. At best, it seems, this idea of the sun as
the heart (his term) of the cosmos was more of a metaphorical model. He put
forward this idea of the sun's rotation being the source of the motions of the
planets about this same time in an unpublished letter, but importantly he calls
it a speculation. Some scholars take this seriously, but I think it shows how far
Galileo had to stretch things in order to support Copernicus within the context
of a biblical passage.

Notes and References

I have used Edward Rosen's translation of Copernicus's *On the Revolutions* (Baltimore:
Johns Hopkins University Press, 1978). Copernicus's value for the precession of Earth
is in Book III, Chapter 6. Also, by Rosen, "Copernicus on the Phases and the Light of
the Planets," *Organon*, 2 (1965), pp. 61–78. Bernard R. Goldstein, "The Pre-Telescopic
Treatment of the Phases and Apparent Size of Venus," *Journal for the History of
Astronomy*, 27 (1996), pp. 1–12.

Richard S. Westfall threw down the gauntlet in, "Science and Patronage: Galileo and the
Telescope," *Isis 76* (1985), pp. 11–30. Coming to Galileo's defense were Owen
Gingerich, "Galileo and the Phases of Venus," published in *Sky and Telescope* (1984)
reprinted in *The Great Copernican Chase and Other Adventures in Astronomical
History* (Cambridge: Cambridge University Press, 1992), pp. 98–104; Stillman Drake,

"Galileo, Kepler, and the phases of Venus," *Journal for the History of Astronomy,* 15 (1984), pp. 198–208; and Paolo Palmieri, "Galileo and the Discovery of the Phases of Venus," *Journal for the History of Astronomy,* 32 (2001), pp. 109–129; for the quotation from Galileo's letter of January 1, 1611 to Kepler, see pp. 119 and 128, n. 30. Gingerich's obituary of Westfall appeared in the *Journal for the History of Astronomy,* 28 (1997), pp. 184–185. The definitive biography of Newton is still Westfall's *Never at Rest: A Biography of Isaac Newton* (Cambridge: Cambridge University Press, 1980).
I first analyzed Galileo's work on sunspots in two articles: "Galileo, Sunspots, and the Motions of the Earth: Redux," *Isis,* 90 (1999), pp. 757–767, and "'I know that what I am saying is rather obscure . . .': On Clarifying a Passage in Galileo's *Dialogue,*" *Centaurus,* 42 (2000), pp. 288–296. The brief critique of my thesis is in Paul R. Mueller, "An Unblemished Success: Galileo's Sunspot Argument in the *Dialogue,*" *Journal for the History of Astronomy,* 31 (2000), pp. 279–299. My defense appeared as, "Colluding with Galileo: On Mueller's Critique of my Analysis of Galileo's Sunspots Argument," *Journal for the History of Astronomy,* 34 (2003), pp. 75–76, which is supported by the following Comment by Owen Gingrich, "The Galileo Sunspot Controversy: Proof and Persuasion," pp. 77–78. I must acknowledge my indebtedness to the pioneering article by A. Mark Smith, "Galileo's Proof for the Earth's Motion from the Movement of Sunspots," *Isis,* 76 (1985), pp. 543–551.
I have used Stillman Drake's translation of Galileo's *Dialogue Concerning the Two Chief World Systems—Ptolemaic and Copernican* (Berkeley: University of California Press, 1967).
For the story of Galileo and the nebulae, see Owen Gingerich, "The Mysterious Nebulae, 1610–1924," published in 1987 in the *Journal of the Royal Astronomical Society of Canada,* and reprinted in *The Great Copernican Chase and Other Adventures in Astronomical History* (Cambridge: Cambridge University Press, 1992), pp. 225–237. Thomas R. Williams presents his case at the Web site of the Galileo Project: http://es.rice.edu/ES/humsoc/Galileo; go to "student work."
I have used Albert Van Helden's translation of Galileo's, *Sidereus Nuncius: or the Sidereal Messenger* (Chicago & London: University of Chicago Press, 1989).
Galileo's discussion of the Book of Joshua is from his letter to the Grand Duchess Christina (1615) reprinted in Stillman Drake, *Discoveries and Opinions of Galileo* (New York: Doubleday, 1957), pp.175–216, esp. 211–215, and Maurice A. Finocchiaro, *The Galileo Affair: A Documentary History* (Berkeley: University of California Press, 1989), pp. 87–118, esp. 114–117. The unpublished letter on the speculated planetary mechanism (to Monsignor Dini, March 23, 1615) is reprinted in Finocchiaro, pp. 60–67, see p. 66. Mario Biagioli seems to take Galileo's speculations seriously in *Galileo's Instruments of Credit: Telescopes, Images, Secrecy* (Chicago: University of Chicago Press, 2006), Chapter 4.

5
Progress Through Error: Stars and Quasars—How Big, How Far?

With the announced discovery of quasars in 1963, a controversy arose around their size and distance. A very similar controversy took place in the Scientific Revolution around the size and distance of the stars. Moreover, at the center of each dispute were two conflicting models of the cosmos.

This story is of even more interest for a quirky reason: as implied in the title of this chapter, the earlier discovery was grounded on a mistake. Did history repeat itself?

5.1. Copernicus and the Distance of Stars

When Copernicus adopted the heliocentric system, the relative distances of the planets were calculated from the geometrical relations among them. This was a direct method of obtaining such distances, unlike the ancient assumption of assuming nesting spheres (see section 3.2 and Table 3.2). The distance Copernicus deduced from the centered sun to Saturn, the last visible planet, was much less than (almost half of) that for the centered Earth to Saturn on Ptolemy's geocentric model. Thus, contrary to what is often said about the difference between the "cozy" medieval cosmos and the modern one, the planetary system of Copernicus was actually smaller. But this was only true for the *planetary* system; when he added the stars, the cosmos expanded indefinitely. He said the entire cosmos was "vast." Why?

In the nesting-spheres model of Ptolemy, the stellar sphere is placed just beyond the sphere of Saturn. Copernicus, however, deduced the distance to the stars by another method, based on the geometry of the heliocentric model. Assuming a moving Earth, then in our semiannual motion around the sun, the stars should visually appear to shift their position by at least the angle 2α (see Fig. 2.4). Moreover, and importantly, the geometry implies that by measuring this angle and knowing the distance between Earth and the sun (the astronomical unit, 1 AU), we can deduce stellar distances. Any such relative shift in an object caused by motion is called parallax; in this case it is stellar parallax, a parallax of a star. But here is the rub: no stellar parallax was observed—not by ancient astronomers, not throughout the Middle Ages and Renaissance, and not by Copernicus. Indeed the absence of stellar parallax was seen as proof that Earth does not move, ever since Aristarchus proposed such an idea, if not before (see section 2.3).

Copernicus, however, turned this "proof" on its head. From Figure 2.4, note that as the distance to the stars increases, the angle α decreases. Mentally extended this distance indefinitely and α approaches zero. So if the stars were extraordinarily far away, stellar parallax would be too small to be seen. This, in essence, was Copernicus's defense. Earth moves around the sun but we don't see the resulting stellar parallax because the stars are extremely far away in comparison with the size of Earth's orbit, resulting in angle α being imperceptible. Hence the cosmos is "vast."

Of course, this is not a proof that Earth moves; on the contrary, in a sense Copernicus was advocating changing the world (specifically, the size of the universe) in order to fit his theory. But the idea did potentially thwart the disproof of heliocentrism. In this way, then, Copernicus enlarged the cosmos, and the idea of an enormous distance to the stars accompanied the heliocentric system.

Astronomers, however, did not initially take up the possible reality of a moving Earth; at most they used his system for making calculations, while ignoring the physical implications. Only some poets and mystics were inclined to exult over a moving Earth in an immense cosmos.

On Being Skeptical About Skepticism

The study of the history of science—and especially the way ideas taken as common sense at one time turn out to be wrong later—may lead one to an extreme form of skepticism, believing that all seemingly definitive knowledge is ultimately fallible. The logic is this: since humans have been shown to be wrong so much of the time, then they are at present probably wrong and will continue to be wrong into the future. A person holding such a position may be called a radical fallibilist.

But this viewpoint is not completely consistent: to be so one should be equally skeptical about being wrong—and thus, accordingly, accept the possibility that humans may indeed be right in many cases, too. In other words, turning things around, our certainly about the fallibility of knowledge may be wrong. This is the standpoint I fancy. Concretely put, maybe many of our present scientific beliefs *will* prove to be true for all time. (Some candidates: the general spherical [really oblate] shape of Earth and its motion around the sun, the path of projectiles, geometrical optics, the periodic table of the elements, the role of DNA in heredity, the measurements of various constants [even if not absolutely constant] of nature.) This position may be categorized as an optimistic form of skepticism, counter to its usually more pessimistic version.

The philosopher Colin McGinn writes of this sensibly optimistic version of fallibilism this way: "Fallibilism says that we cannot be certain what the truth status of our beliefs is, and this leaves room for the possibility that an enormous amount of what we believe is perfectly sound and will never be revised (in fact, I think this is very likely to be the case). A fallibilist cannot consistently maintain that he is *sure* that our current beliefs will eventually be shown false, since he does not think we can be sure of anything. Fallibilists tend to stress the pessimistic possibility, but the optimistic possibility cannot be ruled out, by their own lights."

I think this optimistic version of skepticism and the fallibility of knowledge is more concordant with the history of science, or, at least, my reading of it, as I hope this book demonstrates

5.2. Tycho and Parallax

In the latter years of the 16th century, the most important observational astronomer in Europe was the Dane Tycho Brahe. Initially attracted to astronomy through astrology, he found that the published tables predicting forthcoming celestial events (such as conjunctions of planets) were far off the mark. So he set for himself the goal of correcting these errors. His subsequent meticulous measurements of the heavens, using very large instruments, were the most accurate ever made, surpassing those of Hipparchus in the late ancient world. (The paucity of astronomical measurements from the time of Hipparchus to Tycho is dramatically acknowledged by Owen Gingerich, who reports that less than a dozen observations of the planets are found in manuscripts over this long time period.)

Tycho devoted his life to making precise astronomical measurements, and his strict empirical proclivity precluded giving credence to the reality of a moving Earth. Nevertheless, Tycho could not completely ignore Copernicus's work, since it was being used for astronomical calculations. Moreover, there was the rather elegant deduction of the planetary distances accompanying the model, something even an empiricist like Tycho appreciated (see section 3.3); as he wrote, the heliocentric model "expertly and completely circumvents all that is superfluous or discordant in the system of Ptolemy." But however much its mathematical elegance was attractive, the model's accompanying moving Earth was likewise repugnant, since Earth is a "hulking, lazy body, unfit for motion," as he brusquely put it. Earth must surely remain fixed at the center of the sphere of the fixed stars.

If only Copernicus's mathematical order could be preserved without resorting to a moving Earth, something Copernicus believed was impossible. In key passages in *De Revolutionibus*, on the harmonious nature of the heliocentric system, he speaks of his "arrangement" as displaying "a marvelous symmetry of the universe, and an established harmonious linkage between the motions of the spheres and their sizes." Here Copernicus is referring to the hierarchical order of the relative distances of the planets from the sun, and the corresponding and correlating hierarchical periods of revolution of the planets, something that was not deduced on the

Ptolemaic system (see sections 3.2 and 3.3). Specifically the relative distances of the sun, Mercury, and Venus were unknowable in Ptolemy's model. So Copernicus is right to point to this deduction from his system as a first. And he concludes— this being a continuation of the previous quoted phrase—with this assertion: "Such as can be found in no other way." That is, he thought his arrangement was unique; it was the only way to preserve the mathematical elegance.

In 1588 Tycho put forward an alternative cosmological model, a sort of compromise between the Ptolemaic and Copernican models, in which Earth remained fixed at the center of the stellar sphere, with the sun therefore revolving around Earth, but the planets, in turn, revolved around the moving sun, resulting in an asymmetrical and rather inelegant sort of motion (mentally rotate the spheres in Fig. 5.1 and you'll see what I mean). Tycho preferred this model because, he

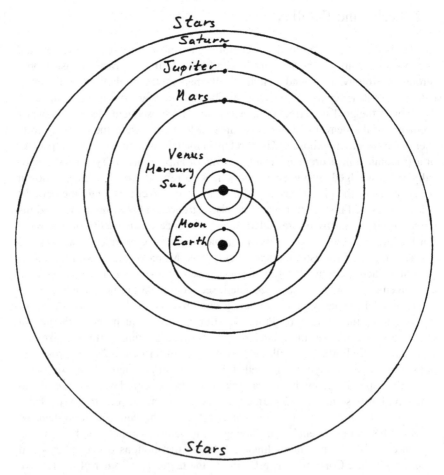

FIGURE 5.1. Tychonic system. Tycho's geo-heliocentric model. The sun revolves around a stationary Earth, while the planets revolve around the sun. Note that it is not drawn to scale.

wrote, it "offended neither the principles of physics [namely, a stationary Earth] nor Holy Scripture."

Tycho's model was offhandedly rejected by those (such as Galileo and Kepler) predisposed toward a more symmetrical sense of order. But there was one important feature of the model of particular importance here: it preserved the mathematical deductions of Copernicus. This is clearly seen in a diagram of 1619 taken from a work by Kepler (Fig. 5.2), who became Tycho's assistant the year before he died in 1601. This is an important diagram in the history of visualization in astronomy, for Kepler shows the relative planetary distances drawn to *scale*, something Copernicus never did (compare Fig. 3.9). (Kepler is the first to draw scaled heliocentric diagrams; another one is found in the *Mysterium Cosmographicum*; see Fig. 6.4.) There are details in this diagram I leave for later (see section 6.3), but note the important element here: the depicted (average) distances from the centered-sun to the planets correspond to the relative distances listed in Table 3.2; see section 3.2). True, the diagram depicts the heliocentric model; indeed, Kepler was perhaps the first astronomer (even before Galileo) to profess publicly his belief in the Copernican system. So how does this relate to Tycho's model? Well, a closer look reveals a dotted circle, centered on Earth and passing through the sun, labeled (translated into English) the "path of Tycho's sun." In other words (and this requires a bit of mental gymnastics), this is really a dual diagram, concurrently exhibiting the Copernican and Tychonic systems, the latter probably in homage to his mentor. (When Tycho was dying, Kepler is said to have promised to adopt Tycho's system.) I believe this is how Kepler interpreted the diagram, for here is what he says: "These speculations about [heliocentric] harmonies also find a place in the hypotheses of Tycho Brahe, because that author has everything else which relates to the arrangements of the bodies and the combination of their motions in common with Copernicus." If Earth is fixed with the sun orbiting us, the Copernican scheme for the ordering of the planets may still be applied. Tycho was thus able to combine the mathematical hierarchy of the Copernican system with, what was for him, an Earth-centered reality. Most importantly, this explicitly shows, contrary to Copernicus's assertion, that the same relative distances of the planets are compatible with *both* systems. Of course, there probably was no way Copernicus could have conceived of this when he penned his phrase lauding what he (erroneously) believed to be the uniqueness of his deduction.

A caveat: Bernard Goldstein of the University of Pittsburgh, supported by Owen Gingerich, has made the case that such an argument is anachronistic. Tycho's system is geocentric and as such the periods of the planets must be viewed from that perspective. As Gingerich writes, "The only way to handle the planetary periods in the Tychonic scheme without being anachronistic is to use the obvious reference frame provided by the earth-sun line. In that case the planetary periods are their Synodic periods, and the numbers are (Mercury to Saturn): 116 days, 584, 780, 399, 378 [see Table 3.1]. The pattern is far from showing a rhythmic [hierarchical] arrangement." (If this is not transparent, try

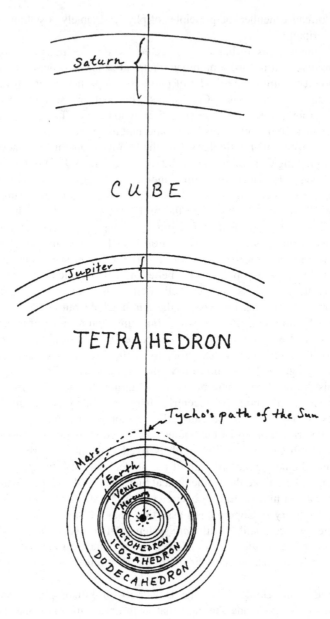

FIGURE 5.2. Kepler's scaled diagram: Copernicus (and Tycho). A sketch of a diagram of the heliocentric model drawn to scale as presented by Kepler in his *Harmonices Mundi* (1619). Not only are the orbits to scale, but the eccentricities of the planarity orbits are shown to scale (showing the range from perihelion to aphelion). These are the result of the elliptical shape of the orbits, first discovered by him. The diagram also includes Kepler's belief that the planets are nested between the Platonic solids (see Figs. 6.3 and 6.4) and his homage to Tycho, with the dotted line around Earth and through the sun, illustrating Tycho's path for the sun on his geo-heliocentric model.

this: Consider Kepler's diagram, Fig. 5.2, at the time of opposition for the Mars, Jupiter, or Saturn. Hence a straight line—vertical in the diagram—connects the sun, Earth, and the planet. The next time they will be lined up again would be the next opposition, and this would constitute a "period" for the planet in this framework. By definition, this is the synodic period. The same argument applies for Mercury or Venus, starting with each lined up with the sun, the planet, and Earth in the vertical line. The synodic period would constitute the time back to the same position, this being one period for the planet in this scheme.) Thus there is no hierarchical sequence of numbers for the periods of the planets in Tycho's model.

True enough. Indeed, this may be one reason that Galileo and others found the model so odious. Nevertheless, it also is an historical fact (witness Kepler's explication above) that the Copernican harmonies are still, so to speak, embedded Tycho's system. It is not only we historians who are finding (or imposing) these numbers within the scheme. Yet I do concede—and furthermore, I hope this chapter emphatically makes the case—that Copernicus's model is aesthetically pleasing—to his, and to my, mind.

A fixed Earth was the cornerstone of Tycho's scheme. And he had further evidence for this geocentric component of his model. Not only was Earth immovable due to its ponderous physical nature, he also convinced himself that he found a way of countering Copernicus's case against the "disproof" of heliocentrism based on the absence of stellar parallax. Let me explain.

Stellar distance is coupled with size. Recall that a dime held at arm's length covers the full moon. Hence the farther away the stars are, the larger in size (volume) they must be (thus we apply Fig. 2.2 to a star). If Copernicus's model is adopted, and accordingly the stars are placed at the minimum distance from the moving Earth such that stellar parallax is not visible, it is possible to estimate roughly a *minimum* size for the stars. Tycho began by measuring the visual angles of the stars (that is the angle of occlusion) and obtained about 2 to 3 minutes of arc for a first magnitude star (2 minutes being just about the observational limit of his very large instruments with the naked eye). Now using the minimum distance to the stars and this angle of occlusion, Tycho was able to estimate their size, because a most interesting deduction neatly follows from Figure 2.4. Think of the eye now placed at the star; from the geometry, the angle of occlusion of Earth's orbit is 2α. Therefore the size of the star must be the size of Earth's orbit—namely, 2 AU in diameter—if the stars are the far enough for the absence of stellar parallax. So Tycho deduced that if the Copernican model were true, then the stars would have to be enormous, as large as (or probably even larger than) the orbit of Earth itself. Unquestionably this was physically impossible; moreover, there would be thousands of these immense objects, at vast distances from Earth, beyond the expansive space past Saturn, which was all too improbable, bordering on the unimaginable. On the other hand, if the stars were closer (as entailed on Tycho's model), they could reasonably and conceivably be smaller and still occlude the same angle. Tycho concluded that this was so. "Hulking, lazy" Earth remained at rest and the stars were still of reasonable sizes.

Tycho believed he had exposed the error in Copernicus's defense against the disproof of heliocentrism, hence undermining Copernicus's argument. Heliocentrism was once more shown to be false, unless, of course, one were willing to accept a reality of vast celestial distances and monstrous stars. Even Galileo (as we shall see) could not fathom the latter.

Tycho's Model in Motion

There were criticisms directed toward Tycho's model from the start. Just looking at the asymmetry of it should raise theological question about God's wonky sense of order. Furthermore, if you conceive of it in actual motion, it presents a rather wobbling display of celestial rotations. (Today we can do more than conceive this motion, thanks to computer animation; see the Web site of Dennis Duke for this and other systems, as cited in the notes.) The major problem at the time was the overlapping of the path of Mars with that of the sun. At first one might conclude that these bodies could collide; a little inspection, however, reveals that that cannot happen, since Mars is always going around the sun. Nevertheless, there was a related problem: because of this overlap, it was deemed impossible for celestial spheres to rotate. As far as I know, this problem was conceived of as fundamentally lethal to the reality of celestial spheres from Tycho onward. Yet, recently, due to the clever insight of psychologist Howard Margolis of the University of Chicago, this deduction has been reassessed. Calling the problem a 400-year illusion, he argues that moving spheres are compatible with Tycho's model by only using spheres for each of the planets, not the sun. All planets then move around the sun on their corresponding spheres while the sun, in turn, moves around Earth, its path thus merely being an imaginary line (such as the equator on Earth). The planets then pull the sun along its prescribed path and there is no need for the sun to have its own sphere. Interestingly, such a model will work, but is it historically relevant? I say no. In the first place, the argument is irrelevant in that Tycho had abandoned the use of spheres. Of course, others may wish to retain the spheres, but even so there is still a problem. Margolis's version must assume that the planets have an internal source of power. But such a model is redundant in the sense that there is then no need for spheres; why not just let the planets move in their predetermined path? More importantly, historically the motions of the planetary-spheres model was based on the assumption that the cause of motion originate at the stellar sphere and this "first cause" of motion was then transmitted between the contiguous spheres from Saturn through Mercury. Such a physical mechanism is not compatible with Tycho's model.

5.3. Galileo on the Stars

In the early years of the 17th century, few astronomers (besides Kepler) believed in the reality of the Copernican system. More were increasingly inclined toward Tycho's system (except for those who felt it violated their sense of symmetry), and by midcentury it was widely used and essentially replaced the Ptolemaic model.

Galileo viewed Tycho's system, with its asymmetrical form and corresponding erratic and off-center motions, as cumbersome and ugly. Perhaps as early at the 1590s, and certainly by the start of the new century, Galileo was thinking more and more from a Copernican viewpoint, especially trying to demonstrate a physics of motion that would "work" on a moving Earth (see section 8.3). When Kepler published his affirmation of the heliocentric model, Galileo (in a letter to Kepler) praised his courage. But he himself did not profess it in print.

It was not until his telescopic sightings that Galileo seems to have become fully convinced of the truth of the Copernican system, which he made public in *Sidereus Nuncius* (1610), his book on the celestial revelations perceived through his telescope. Of the many discoveries disclosed in that small but wonderful book, one is specifically germane to the topic here. In looking at individual stars, Galileo notes that they appear, not as uniform disks, but with distinct haloes around them. In his words, "When the stars are observed with the naked eye, they do not show themselves according to their simple and, so to speak, naked size, but rather surrounded by a certain brightness and crowned by twinkling rays. . . . Because of this they appear much larger than if they were stripped of these extraneous rays, for the visual angle is determined not by the primary body of the star but by the widely surrounding brilliance." So the visual angle does not measure only the physical star, but the star plus its halo; measuring only the *physical* size, Galileo found a first magnitude star's diameter to be about 2 seconds of arc, about 1/24 or 1/36 of Tycho's determination. Thus, the actual star is really much smaller in diameter, and hence in volume too, than previously thought.

This had immediate bearing on Tycho's "disproof" of Copernicus. The stars need not be of enormous size to be at the required ("vast") distance for the absence of stellar parallax. The physical basis of Tycho's argument, therefore, is discredited and the disproof collapses. Copernicus's system is no longer falsified and thus it remains a viable model of reality.

This argument, however, does not constitute a proof of the Copernican system. To be logically consistent, it only eliminates a disproof of the Copernican model; proof would require further independent confirmation. Tycho's system, for example, is still compatible with Galileo's discovery. Thus both systems are back on equal empirical footing, so to speak.

What is most fascinating about this entire controversy is that it was based on a commonly accepted "fact," which we now know to be dead wrong, namely, that the stars cannot be of enormous size. Neither scientist questioned this assumption; both Tycho and Galileo agreed on the physical impossibility of an object the size of the orbit of Earth. In his *Dialogue* (1632), Galileo put the idea in the month of Simplicio, the interlocutor holding to the geocentric model, this way: "Such bulks are truly too vast, and are incomprehensible and unbelievable." This was surely the position of Galileo, too, since the Copernican speaker never questions this "fact."

Today we know, and freely accept, the reality of stars even larger than this. Of course, we know that Galileo was ultimately right about the Copernican system, and hence his disproof of Tycho was an important step toward the ultimate acceptance of the Copernican system. Yet along this progressive path was an erroneous assumption about the physical nature of the stars. Hence an error was an agent in what in the end was a correct deduction. How ironic!

(We know today there was a deeper error: the halos were diffraction rings and that Galileo was not measuring stellar diameters, which were actually about 100 times smaller than his 2 seconds of arc.)

Stellar Distances and Parallax

Galileo's telescope revealed that there were many more stars in the sky than visible to the naked eye. Where were these other stars? There were two possibilities: either they were on the stellar sphere but much smaller than the other stars, or there was no stellar sphere and they were farther away and hence just appeared smaller. Both deductions were logically possible on the Copernican system but not the Ptolemaic system.

On the geocentric system, where the stars and everything else in the heavens must rotate once a day, the only logical framework is the stellar sphere. Assuming the stars to be distributed in space but still rotating around us so as to remain "fixed" with respect to each other would entail an extraordinarily complex celestial mechanism that adjusted their motions such that their distances and rotational speeds were correlated. The model of the stellar sphere is very much a simpler one.

When Copernicus postulated a rotating Earth to account for the daily rotations of the heavens, then the fixed stars were literally fixed, not only with respect to each other, but with respect to the fixed central sun. This meant that it is logically possible to assume that the stars are not attached to a sphere but that they may extend into space, and without postulating a complex mechanism for their daily rotation (which was taken up by Earth). Copernicus could have made this assumption, but he did not; instead, he elected to keep the stellar sphere in his system. (Incidentally, Kepler too remained committed to the stellar sphere.) But others who adopted the heliocentric model, such as the mystic Giordano Bruno, freely seized on the idea of the stars extending into space, and espoused an infinite universe. This led to the speculation that our sun was just another star and that the stars were really other possible planetary "worlds" in an infinite universe, which was surely science fiction for the times.

Thus when Galileo viewed numerous additional stars beyond those amenable to naked eye observation, the possibly arose that these other stars were farther away than the visible ones, thus supporting the elimination of the stellar sphere. What was his position on this? In the *Dialogue* he has the Copernican say, "I do not believe that the stars are spread over a spherical surface at equal distances from one center; I suppose their distances from us to vary so much that some are two or three times as remote as others." Indeed, from this he proposes a way of measuring stellar parallax, using a more distant star as a reference point to measure the motion of a nearer one. Nevertheless, the fundamental conceptual framework of the Copernicans system, as presented in the *Dialogue*, is based on circularity, both for terrestrial and celestial motions, as I show in some detail (see section 8.4). A stellar sphere, accordingly and fittingly, complements this, providing the endmost circular framework holding the universe together. In a letter he once wrote, "It is as yet undecided (and I believe that it will ever be so for human knowledge) whether the universe is finite or, on the contrary, infinite." In the end, it seems, he remained ambivalent.

Ultimately, through the works of especially René Descartes and Isaac Newton, the infinite universe prevailed. By the 18th century it was the accepted picture of the cosmos, along with the Copernican system for the planets. It was not until 1838 that the mathematician-astronomer Friedrich Bessel discovered a small parallactic shift (0.3 seconds of arc) in the star 61 Cygni, in the constellation Cygnus (the swan). Having found parallax, it was now possible to calculate directly stellar distances by measuring the parallax of a star against the background of the other fixed stars (measuring α and knowing 1 AU in Fig. 2.4, the star's distance is deduced). The star 61 Cygni turned out to be about 11 light-years away (using the distance-measurement scale later adopted). Over the course of the 19th century, further parallactic shifts were measured and the celestial distances increased greatly. Science fiction was transformed into factual science as Copernicus's "vast" space was at last being surveyed.

5.4. Quasars and Cosmology

There is a most interesting parallel between the debate over the sizes and distances of the stars from Tycho to Galileo and one that took place in the 1960s and after, with the discovery of quasars. First appearing on radio telescopes as point-like (and therefore star-like) sources of radio energy, they were initially called "quasi-stellar objects," from which the shorter term *quasar* came. Later they were seen through optical telescopes too. What struck (and even dazzled) astronomers was the fact that quasars seemed to embody an enormous amount of energy (comparable to millions of stars) in a relatively small space (like a star). With so much energy, they could not be stars. But they were not galaxies either, for it was soon noticed that they changed in brightness over short periods of time (from weeks to months), and it is impossible for a galaxy to pulse this way (for example, it takes about 100,000 years for light to get from one end of our galaxy to the other, and our Milky Way is of modest size). In other words, these objects seemed to possess the energy of a galaxy within about the volume of a star. It was as if an archaeologist found a fossil in an unsuspected place, and the fossil did not fitting anywhere in the taxonomic system. Totally unexpected and unpredicted, the first quasars were announced in 1963.

A personal anecdote: I was an undergraduate student majoring in physics and mathematics at the time of this discovery, and I can still recall the hoopla about quasars. I attended a lecture on these strange quasi-stellar objects by an undoubtedly eminent visiting physicist/astronomer whose name I don't recall, but I do remember that he proposed a mechanism for their energy based somehow on gravitational forces. From the lecture and the ensuing discussion, I had the sense that the scientists were merely groping for an answer.

Subsequently dozens more and ultimately thousands of quasars were found over the years, along with an important discovery: their light shifted toward the red end of the spectrum, exhibiting redshift. What did this mean?

By the early1930s Edwin Hubble and his assistant, Milton Humason, working at Mt. Wilson observatory, had shown that the light from most of the galaxies exhibited redshift (although they were not the first to observe this) and importantly they found a correlation between this redshift and the distances of the galaxies. (Hubble really spoke of nebulae, because the distinction that we use today between nebulae and galaxies was not yet made.) The result was a nearly linear relationship between the redshifts and the distances to these galaxies (see section 12.3). As to the origin of the redshift, one candidate was the Doppler shift. Although this analogy between light and sound was questioned in the 19th century, the optical Doppler effect was reinforced by Einstein's relativity theory in 1905. Not only did he postulate that the speed of light was independent of the motion of the source (as was true for sound waves), but (not often recognized) he deduced from electromagnetic theory a relativistic (optical) Doppler effect. As relativity was eventually embraced by the community of physicists, so was the optical Doppler principle. That did not mean, however, even as late as the 1930s, that the redshifts of the galaxies were indeed identified as Doppler shifts. Not only Hubble had doubts; the astronomer and physicist Fritz Zwicky at the California Institute of Technology admitted the validity of the optical Doppler effect locally, but questioned applying it to the total universe. Recognizing that light is bent by gravity in general relativity (see section 1.1), Zwicky made the argument that light, as it travels though space at a finite speed over eons of time, may be retarded as it recurrently passes and is bent by large masses; he called this "the gravitational drag of light" (it was later called by others the "tired-light" hypothesis). Zwicky proposed another interesting objection (interesting from today's viewpoint, I think): since Einstein had not completed the unified field theory (see Chapter 1), then general relativity is incomplete, perhaps only locally valid, and therefore any deductions about the total universe are uncertain. (Well, the unified theory is still incomplete today and yet general relativity forms the foundation of present-day cosmology.)

For those others in the early 1930s who accepted the redshifts of the galaxies as Doppler shifts, this redshift was a measure of recessional velocity, a result eventually called Hubble's law (see Fig. 12.2). Written as, $v = Hd$, it is a simple linear relationship, where v is velocity and d is distance, with the constant H (after Hubble). Moreover, it followed from this that the universe (conceived of as composed mostly of galaxies) would be expanding in time, a deduction that fit into one of the variations of Einstein's theory of general relativity (see section 12.4). In the 1930s, therefore, there was a loose consensus among cosmologists holding to this model of an expanding universe but not without some self-doubt. One key source of doubt was the contradiction between a deduction from the model and an apparent fact from another branch of physics. In Hubble's law, $1/H$ has the unit of time, and hence it is possible to deduce the age of the universe— that is, the time of expansion—from the law. The result was under 2 billion years, which may seem to be a reasonably large number. But it was not large enough, since physicists using radioactive decay to date the age of Earth were getting numbers around 3 to 4 billion years. Earth being older than the universe would be analogous to you being older than your mother. The result was a deadlock: two branches of physics seemed to contradict each other. The doubts about the redshifts by Hubble and Zwicky, among others, were thus justified.

In 1946, three young Cambridge scientists attended a ghost movie titled *Dead of Night,* in which five stories were linked such that the film's beginning became the end; time was in a loop. Supposedly inspired by this, they published in 1948 a paper boldly putting forward an alternative cosmological model, chiefly to resolve the contradiction in time scale entailed in the expanding model. Named the steady-state model, it assumed no beginning or end to the universe (a view, incidentally, that goes back, at least, to Aristotle). Matter is minutely but continually being created in the space left by the receding galaxies, so that over a long time scale, the universe looks essentially the same. This solved the problem of Earth's age versus that of the universe; it also eliminated the need to postulate a beginning—an extra-scientific feature of the expanding model, which some viewed as more theological than scientific. Finally, in a display of rather sophomoric exuberance, one of them bestowed on the rejected expanding model a derogatory name—the Big Bang model, an appellation that unfortunately stuck. The term is unfortunate in another way (independently of sexual innuendos): the "bang" suggests an explosion at the beginning of the universe (a popular misconception). The model, however, implies no such thing; instead, in the beginning all the energy (and space and time) is compacted together, after which the expansion proceeds as space and time arise concurrently with energy converting to mass ($E = mc^2$)—all this going on continuously. An analogue: a balloon blowing up does not begin with a bang (although it may end as such).

In the 1950s cosmologists had a choice between these two universes. Both the steady-state and the Big Bang models had staunch adherents throughout the decade and into the next, even after the contradiction in time scale was eliminated. In the mid-1950s, using data mainly from the large telescopes high in the California mountains (e.g., Mt. Palomar), Hubble's law was recalibrated, with a corresponding change in the value of H; in Figure 12.2, the slope decreased. As a result the universe's age on the Big Bang model increased considerably, to 10 billion years and more. At that time Earth's age was measured at about 4.5 billion years, and so the contradiction was no more. Nonetheless, for mainly conceptual, aesthetic, and perhaps even theological reasons, many cosmologists still found the steady-state model appealing.

Incidentally, Hubble died in 1953, just before the time scale was extended. A study of his writings from the 1930s to his death indicates that he never accepted the redshift of the nebulae as a Doppler shift, and hence he rejected the expanding universe. As a confirmed empiricist (rather like Tycho), he felt the model went beyond the data. We can only speculate about what he would have thought after the contradiction was removed.

Which brings me back to quasars. When they were discovered, the conflict between the two models came to the fore, since quasars exhibited redshift. Indeed, their redshifts were extremely large (at first about one-fourth the speed of light; later some were found close to the speed of light); using Hubble's law, this placed them at extreme distances, perhaps near the creation of the universe on the Big Bang model. Quasars surely were strange objects of tremendous energy, compact and pulsing, perhaps over 10 billion light-years away—and still visible, no less. Predictably, the question arose: How can such gigantic objects, which seem to defy the present laws of physics, even exist?

Sound familiar? Indeed: this question is an analogue to that put forward by Tycho on the nature of the stars on the Copernican model. His answer was that the stars are in fact much closer and hence not so large as to defy physical reality, and accordingly, Earth was at rest at the center of the cosmos. So, in the 1960s an answer to the problem of the quasars was, likewise, that quasars are much closer and hence need not require so much energy as to contradict the present physical laws. This also meant that redshifts were not Doppler shifts. Such a viewpoint contradicted the Big Bang model but was compatible with the steady-state model. This tilted the evidence in favor of the steady-state model, although the issue of explaining redshift was a liability. Hypotheses to explain away redshift were put forward, such as assuming redshift in quasars was due to internal forces (maybe gravitation) "stretching" the light toward the red end of the spectrum; in any case, this assumption was seen as less a "stretch" of the imagination than the enormous energy required for the quasars being 10 to 15 billion light-years away. Does this also not sound familiar? Recall, for example, the arguments of Zwicky in the 1930s to get around the problem of redshift. By the mid-1960s it seemed that the steady-state model had won the day.

Then in 1965 another empirical discovery was made. Like the discovery of quasars, this too was unexpected, but as it turned out, this one was not unpredictable. Throughout space, in all directions, it was discovered that the cosmos seemed to be filled with microwave radiation. An antenna tracking signals from satellites detected this radiation as background noise. The measured temperature ($-273°$ C) was found to fit the Big Bang model as residual energy "left over" after billions of year. This radiation (sometimes erroneously referred to as an echo of the Big Bang) was predictable (in the sense that it fit previous calculations of the Big Bang model made in the 1940s), although no one had thought to follow it up by actually looking for it! So the discovery was made, by accident. It became known as cosmic background radiation. Because of the mathematical fit, there was much incentive to establish it as evidence for the Big Bang model.

In time the Big Bang was eventually the dominant model, but its acceptance was not immediate, contrary to what is often proclaimed in textbooks. Well into the 1970s and later, there were still sundry astronomers questioning whether redshifts are Doppler shifts and staunchly supporting the steady-state model. It seems that it was not until around the 1980s that a consensus was reached among astronomers supporting the Big Bang model, with cosmic background radiation as the main evidence, and that those still holding to the steady-state model were increasingly viewed as retroactive cranks. In 1980, for example, in a summary article written for the journal *Science,* the celebrated astronomer, Vera Rubin, who, incidentally, was the first woman permitted to use the great telescope on Mt. Palomar, felt confident enough to write, "Most astronomers accept as a model a universe which has expanded and cooled from an initial hot, dense state." At most she made a passing reference to the alternative model. In a subsequent issue of the journal, a letter appeared from a supporter of the steady-state model doubting her assertion that "most astronomers" accepted the Big Bang model and the corresponding explication of redshift. But this was increasingly a minority view, and by the mid-1980s only a few established astronomers clung to the steady-state model.

An aside: I wish to point out that the late Fred Hoyle, one of the creators of the steady-state model, although considered wrong on this today, was right about something else; namely, his postulate, with others, that the heavier elements in the periodic table were formed by the nuclear processes in supernovae.

So what of the analogy with Tycho and Galileo? Can we carry it any further, into the story of quasars? *Prima facie,* it breaks down; although the two stories are parallel, history does not seem to be repeating itself. The shift to the Big Bang model was not the result of a discovery about the size of the quasars, as happened with the conversion to Copernicus. Rather, it seems that, supported by cosmic background radiation, the Big Bang model rode roughshod over the steady-state model and won the day by a slow inculcation. But the problem of the energy of the quasars remained, and I think this was one reason why the adoption of the Big Bang took so long, despite the evidence from cosmic background radiation, so that those supporting the model fumbled around to devise mechanisms to explain it. (And there have been, and continue to be, many explanations: the latest, as I write this, is that quasars are the result of Black Holes sucking stars from the center of galaxies, with the stars giving off intense energy due to the conversion of mass to energy as they accelerate into the Black Hole.)

Is this where the parallel ends? So far no error that we know of has led to progress apropos the place of quasars in the universe. But we are too close to the events to see things clearly. The issue of the source of the quasar's energy needs to be settled before we can dismiss the historical parallel entirely. I believe the drama is not over.

More Thoughts on Skepticism

One dominant mode of skepticism in Western thought is the humanist thread based on logic and epistemology from the pre-Socratics to the present, associated primarily today with academic philosophy: the ideas of Descartes, Hume, Kant, et al. Another variation, but not always independent of the dominant mode, I call theological skepticism; it is based on the Judeo-Christian premise that God's knowledge is infinite and infallible, whereas human knowledge is finite and subject to doubt, however certain we may think we are about something. Pope Urban VIII expressed this humility and intellectual diffidence in the face of God in his famous meeting with Galileo (when they were still friends), requesting that Galileo put this theological maxim in his book. Galileo did so, only he placed it at the end of the *Dialogue* (1632) and in the mouth of Simplicio, the discredited spokesman for Aristotle and Ptolemy. The insult, in part, led to Galileo's trial before the Inquisition.

An interesting modern variant of skepticism comes from Charles Darwin. Of course, he was no skeptic when it came to evolution, but he did express self-doubt about his certainty on things theological. It appears in his short autobiographical essay where, appropriately, he is discussing the existence of God. His argument is this: since humans evolved from primates, we are still linked with them and their primitive ways; we have, in other words, not yet evolved very far, and hence we cannot be fully secure in our knowledge of things. Darwin explains his doubt this way: "Can the mind of man, which has, as I fully believe, been developed from a mind as low as that possessed by the lowest animal, be trusted when it draws such grand [theological] conclusions?" The parallel with theological skepticism is most interesting; the human limit with respect to the Divine (below the angels, above the beast) is now replaced with the human limit in light of its "recent" evolution from (and hence not far above) the lower creatures.

I would add one more version: my own, which I call historical skepticism. The meaning should be quite clear for anyone reading this book. The study of history reveals the vicissitudes of certainty, knowledge, and belief, and thus inexorably leads toward a healthy (I believe) skepticism about most things, tempered, however, with my own doubt about my doubting (see On Being Skeptical About Skepticism, section 5.1).

Notes and References

I wrote on some of the topics of this chapter in David R. Topper, "20th Century Quasars and 17th Century Stars," *The Physics Teacher,* 13, No. 6 (September, 1975), pp. 339–342.

The quotations from Tycho Brahe are taken from Owen Gingerich and James R. Voelkel, "Tycho Brahe's Copernican Campaign," *Journal for the History of Astronomy* 29 (1998), pp. 1–34 (on pp. 1 and 23–4). See also Bernard R. Goldstein, "Copernicus and the Origin of his Heliocentric System," *Journal for the History of Astronomy* 33 (2002), pp. 219–235. Owen Gingerich's defense of Goldstein is from an email to the History of Astronomy Listserv Group (July 20, 2005).

The Web site of Dennis Duke (Physics, Florida State University), displays animations of Tycho's and other astronomical systems: http://www.csit.fsu.edu/~dduke/models. Howard Margolis presented his argument in "Tycho's Illusion and Human Cognition," *Nature* 392 (April 30, 1998), p. 857. A debate (involving others and me) over his thesis is found in the electronic journal, *Psycholoquy,* 1998: H. Margolis, "Tycho's Illusion: How it Lasted 400 Years and What that Implies About Human Cognition," *Psycholoquy* 9 (32); D. Topper, "Margolis's Delusion: A Critique of 'Tycho's Illusion,'" *Psycholoquy* 9 (42), and "Struggling with Tycho's Spheres: A Retraction and a Supplement," *Psycholoquy* 9(61).

Galileo, *Sidereus Nuncius: or the Sidereal Messenger,* Albert Van Helden, trans. (Chicago and London: University of Chicago Press, 1989). Galileo, *Dialogue Concerning the Two Chief World Systems—Ptolemaic and Copernican,* Stillman Drake, trans. (Berkeley: University of California Press, 1967). Copernicus, *On the Revolutions,* Edward Rosen, trans. (Baltimore: Johns Hopkins University Press, 1978). Galileo's letter on stellar distances is quoted in Alexandre Koyré, *From the Closed World to the Infinite Universe* (Baltimore and London: Johns Hopkins University Press, 1957), p. 97.

Vera C. Rubin, "Stars, Galaxies, Cosmos: The Past Decade, the Next Decade," *Science,* 209, No. 4452 (Centennial Issue: July 4, 1980), pp. 63–71. I have used two papers by Fritz Zwicky, "On the Red Shift of Spectral Lines through Interstellar Space," *Proceedings of the National Academy of Sciences* 15 (1929), pp. 773–779, and "Remarks on the Redshift from Nebulae," *Physical Review* 48 (November 15, 1935), pp. 802–806. See also Stephen G. Brush, "How Cosmology Became a Science," *Scientific American,* 267, No. 2 (August, 1992), pp. 62–70; "Prediction and Theory Evaluation: Cosmic Microwaves and the Revival of the Big Bang," *Perspectives on Science,* 1, No.4 (Winter, 1993), pp. 565–602; and "Is the Earth Too Old? The Impact of Geochronology on Cosmology, 1929–1952," in C.L.E. Lewis and S.J. Knell, eds., *The Age of the Earth: from 4004 BC to AD 2002* (London: Geological Society, 2001), pp. 157–175.

My remarks on skepticism were inspired by two paragraphs written by Colin McGinn (professor of philosophy at Rutgers University) in his review of the book, *Invariances: The Structure of the Objective World,* by Robert Nozick, in *The New York Review of Books* (June 27, 2002), pp. 39–41, on p. 39. I have used *The Autobiography of Charles Darwin, 1809–1882,* edited by Nora Barlow (New York and London: W.W. Norton, 1969).

6
The Data Fit the Model
but the Model is Wrong: Kepler
and the Structure of the Cosmos

I may be going out on a limb, but I'm convinced that science often is daydreaming constrained by reality (and logic). I hope this chapter convinces you of this outrageously sounding proposition.

6.1. Neutrinos and Sydney's Opera House

When I was a student in the 1960s reading art history, I came across something I found odd. Pictures of the Sydney Opera House in Australia were always reproduced as drawings or scale models, not photographs. A little research quickly explained why: the building did not exist. The following musings are grounded on that discovery.

In 1930 a problem arose in physics centered on experiments involving something called beta decay. Beta particles were really fast-moving electrons emitted by the nuclei of radioactive material. The problem with beta decay was that the energy before and after the decay was not the same. This seemed to contradict the law of the conservation of energy. Established in the 19th century, the law asserted that during all processes in nature (such as, say, transforming water into steam), the energy before and after the change was the same. The law was modified in the early 20th century to accommodate Einstein's $E = mc^2$, but otherwise remained one of the foundations of all physical as well as chemical processes. How could beta decay be explained without violating a law of nature?

The physicist Wolfgang Pauli proposed a solution. It came in the form of a letter sent to a conference on radioactivity, a conference that he did not attend because he was needed—he said he was "indispensable"—at a ball. He referred to his solution as "a desperate way out" of the problem. Briefly put, he postulated the existence of an as-yet-unknown particle that carried just the right amount of energy to balance the conservation equation. At the time only three elementary particles were known to exist: the electron, the proton, and the photon (or quantum of light). The radical nature of this postulate for the time is revealed by the reaction of Niels Bohr, who had put forward the quantum explanation of the atom in 1913. Bohr was more comfortable abandoning the energy conservation law in nuclear processes than in adding another particle to the physical world.

Pauli originally called his particle a "neutron," since it entailed only a very small quantity of mass (and therefore energy); otherwise it was neutral, that is, without electrical charge, unlike the electron (negative) and proton (positive). But by 1932 the term was usurped when the "real" neutron was discovered, and hence another word was coined for Pauli's particle: Enrico Fermi called it a "neutrino," from an Italian word for a little neutral object.

Of more than passing interest are the tumultuous events in Pauli's life at the time of his "desperate" solution. He was distressed by the recent suicide of his mother, he had divorced his wife several days before writing the letter, and he was drinking heavily. Two months before he died in 1958 he recalled the now-famous letter, and called the neutrino "that foolish child of the crises in my life." Foolish or not, ultimately the neutrino was eventually found in nature, providing a Nobel Prize for two experimental physicists, Fred Reines and Clyde Cowan. Today talk of neutrinos is as commonplace among physicists as electrons and protons, along with quarks and other fundamental things. But—and here is the crux of the matter—the experimental confirmation, which was transmitted by telegraph from Reines and Cowan to Pauli ("We are happy to inform you that we have definitely detected neutrinos"), is dated June 14, 1956, that is, 26 years after Pauli postulated his "foolish child." In the meantime the neutrino existed only as an entity in scientific writings; nevertheless, its eventual physical being was seldom doubted because few scientists (beyond Bohr) were willing to tinker with the law of energy conservation. Over those years, the neutrino existed, but only on paper.

In September 1955 an architectural competition was announced in the city of Sydney for designs for a center for the performing arts. In December 1956, about six months after the discovery of the neutrino, the winner of the competition was announced, Jørn Utzon of Denmark. From the start it was a controversial decision; the design seemed frivolous, disorderly, and unworkable. Yet Utzon insisted there was an underlying order to the proposed arching vaults, that they were, in fact, individual sections of a sphere, he claimed. Ground was broken in 1959 but the work was not completed until late in 1972 with the official opening in October 1973—almost 17 years after Utzon received the commission. (There were numerous delays and setbacks. Utzon quit at one point. Indeed, the project was almost terminated several times.) Throughout that period the existence of the Center for the Performing Arts (later shortened to the Sydney Opera House) was confined to art books, as I discovered sometime in the 1960s.

Both the neutrino and the opera house existed on paper between their conceptualizations and eventual realizations for the same reason—some imagery is so good that it must exist. A conservation law in science is essentially a law of harmony, a balance between two entities; the latent neutrino preserved that equilibrium. The potential opera house was a marvelous example of an antifunctional piece of contemporary architecture suited for inclusion in an art book (Sigfried

Giedion, in the 1967 edition of his classic, *Space, Time, and Architecture,* devoted over 15 pages to it).

And this leads me to the main story of this chapter, on Kepler and the structure of the cosmos.

6.2. Kepler and God's Mind

In March 1610, Johannes Kepler, "Imperial Mathematician" to Rudolf II in Prague, received a report that Galileo discovered four new "planets" using a "perspicillum" (later called a telescope, note its etymological connection with perspective). Kepler was sure these "planets" were not objects on a par with Venus, Jupiter, and the like, but rather were smaller "moons" circling the planets, like our moon. (As an aside, Kepler later coined the term "satellite.") Being a Copernican and therefore thinking of the universe as homogeneous, he presumed that Galileo had discovered a moon around each of the planets, from Venus through Saturn; he further surmised—Kepler was prone to excessive speculation on meager data— that Galileo did not yet see the moon of Mercury because the planet is so close to the sun. Later, after obtaining a copy of Galileo's *Sidereus Nuncius,* Kepler learned that he was correct in his deduction of the "planets" actually being moons; however, Galileo had found that all four moons circled Jupiter alone. Ah ha! This sent Kepler now thinking this way: since Earth has one moon and Jupiter four, then probably Mars has two and Saturn has either six or eight; as well, Mercury and Venus have one each. To be sure, Kepler was thinking in terms of sequences: 1, 1, 1, 2, 4, 6 or 1, 1, 1, 2, 4, 8. Nature was ordered; in particular, this order was mathematical (either arithmetical or geometrical), having its ultimate origin in the mind of God. Kepler's science, as we will see, was a branch of theology.

Interestingly, Kepler's prediction was partially correct: Mars does have two moons, which (being so small) were not discovered until the late-19th century. Moreover, so far as we know, it has only these two (Demos and Phoebus), whereas Jupiter and Saturn have several dozen each, as we have found in recent years.

Let us back up about $1\frac{1}{2}$ decades to Kepler's first work in astronomy. As a student at the University of Tübingen in Germany, Kepler was enrolled to study theology with the goal of entering the Lutheran church. But he also excelled in mathematics, and at about the time he obtained his master's degree, a teaching position in mathematics (which included astronomy) opened at an affiliated seminary in Graz, Austria. Kepler was recommended for the job since he was the best-qualified student, but he balked at the offer because astronomy was not his chief interest. As well, he said it was his worst subject; he received an A−, whereas all his other grades were As! Nevertheless, he was eventually persuaded to accept the job.

The move sealed his fate. Although his duties at Graz were more astrological than astronomical (he cast horoscopes to predict the weather and the like), they were not onerous, and he had much spare time in which to think—and think he did, about astronomy. The astronomy he was asked to teach was, of course, geocentric—Ptolemy's system. But Kepler's teacher at Tübingen had exposed him to Copernicus's heliocentric model, and he was especially attracted to it. So the astronomy he was contemplating (but not teaching) was Copernican, and this deliberation eventually led to his first book, published with money out of his own pocket.

The key idea in the book grew out of his belief in a mathematical order of the universe coupled with a "fortuitous" (Kepler's word) event in July 1595. Most of that summer Kepler had been pondering this question: Why are the planets at their relative distances from the sun (based on, of course, the Copernican system)? To answer this he was looking for some pattern to the numerical sequence for the planets from Mercury to Saturn, setting Earth–sun distance to 1 AU: 0.36, 0.72, 1, 1.5, 5, 9 (see section 3.2 and Table 3.2). He assumed there was an arithmetical law embedded in it (as we saw he would later do with Galileo's moons). But nothing worked and he "wasted" (he tells us) the summer on it. Then, he writes, "I thought it was by divine intervention that I gained fortuitously what I was never able to obtain by any amount of toil; and I believed that all the more because I had always prayed to God that if Copernicus had told the truth things should proceed in this way." The latter clause is crucial to understanding the mind of Kepler. As a way of testing the validity of the Copernican model, he made a wager with God, who would give him a sign if heliocentrism were true.

The sign came in the form of a geometrical diagram. While teaching a class on the conjunctions of the planets Saturn and Jupiter, which occur almost every 20 years, Kepler drew a diagram showing where the conjunctions occur along the zodiac, namely a little less than 240° apart. (Since Saturn's period is 30 years, and 20/30 = 240/360, then each conjunction occurs along the Zodiac almost every 240°.) If it were exactly 20 years, the result would be an equilateral triangle. But being a bit less than 20 years, there results a series of quasi-triangles as in Figure 6.1. What struck Kepler about this diagram was the approximate circle that was inscribed by the quasi-triangular lines.

This led him to another image: if an equilateral triangle is inscribed in a circle and another circle within it (Fig. 6.2), then the ratio of the radii of the two circles is fixed at 2 to 1 for all cases, independently of the absolute sizes of the circles. (I leave it for the reader to prove this; hint, think of similar triangles.) Now 2 to 1 is close to the ratio of the distances of Saturn and Jupiter, namely 9 to 5. Perhaps God used such a geometrical pattern for fitting the planets in their orbits? Kepler's mind, as always, raced ahead: so next would be Jupiter to Mars, and the fitting of a square; then Mars to Earth, and a pentagon, and so forth to Venus and Mercury, with corresponding polygons within the spacings. When he tried fitting a square between Jupiter and Mars, the ratio of the radii of the outer to inner circles (1 to $1/\sqrt{2}$; or 1 to 0.707, setting the outer radius to unity) was nowhere near the predicted value (5 to 1.5; 1 to 0.333). He tried combinations of polygons (e.g., a square within a triangle), but this didn't work, either; nor did adding pentagons, and, hence, after much toil, he gave up on this "useless attempt" to penetrate the mind of God. Neither arithmetic nor geometry seemed to unveil a cosmic pattern.

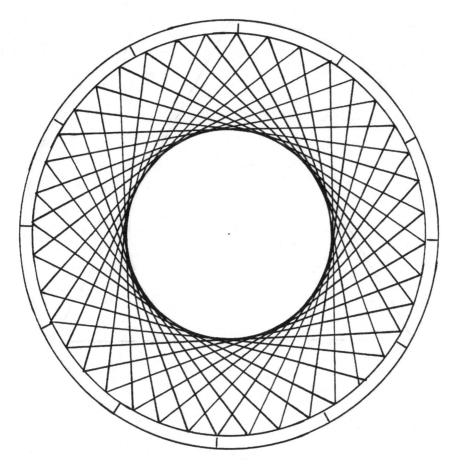

FIGURE 6.1. Kepler: the conjunctions of Jupiter and Saturn. A sketch of Kepler's geometrical diagram of the conjunctions of Jupiter and Saturn from which he conceived of the nesting of the spheres between geometrical forms (see Fig. 6.2).

Just as he was about to abandon the entire project, something clicked. He realized that not only was he searching for the reason for the distances of the planets, but a parallel question arose: Why are there just six planets, "rather that twenty or a hundred"? This pattern of circles within polygons implied that, using one polygon between each planet, the total pattern required only *five* figures. Moreover, although he had been trying to fit polygons to circles, for the planetary system itself the mathematical forms were really three-dimensional, or as Kepler put it, "solid bodies" between "solid spheres." The idea of five solid forms, "for anyone having a slight acquaintance with geometry," Kepler reminds us, "there would immediately come to his mind the five regular solids." Of course, the so-called five Platonic solids! As Euclid proved, these are the only regular (symmetrical) solids that are formed by using just one polygon for each face (Fig. 6.3). Since

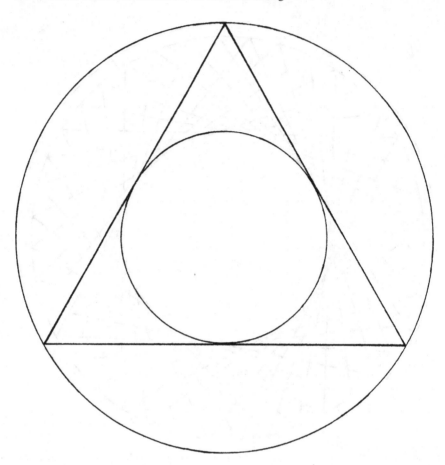

FIGURE 6.2. Kepler: fitting polygons between the planets. An example of the concept of nesting planets, in this case in 2D. Using an equilateral triangle results in a unique ratio for the sizes of the circles. Nonetheless, the ratios of 2D polygons did not fit the planetary data, so Kepler tried 3D forms.

ancient times these five forms were symbols of perfection. Kepler reports he then "made a vow to Almighty God" that if this scheme worked, he would write a book ("I would proclaim among men in public print this wonderful example of his wisdom"). And it did work! It was as if he had found the answer to the question as to "what hooks the sky is hung on to prevent it from falling," as he metaphorically put it.

And so it was that Kepler published his first book, in 1596: the complete title (in English) being, "Forerunner of the Cosmological Essays, Which Contains the Secret of the Universe (*Mysterium Cosmographicum*); On the Marvelous Proportion of the Celestial Spheres, and on the True and Particular Causes of the Number, Size, and Periodic Motions of the Heavens; Established by Means

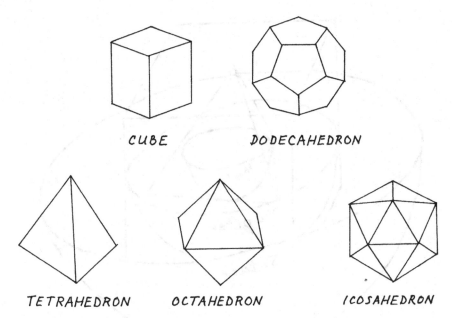

CUBE   DODECAHEDRON

TETRAHEDRON   OCTAHEDRON   ICOSAHEDRON

FIGURE 6.3. The five platonic solids. The 3D forms eventually used by Kepler, the Platonic solids. These are the only 3D forms that have a unique polygon for each face and are symmetrical. They are the cube (or hexahedron: i.e., having six units or sides), dodecahedron (12 sides), tetrahedron (four sides), octahedron (eight sides), and icosahedron (20 sides).

of the Five Regular Geometric Solids." Historians simply call it *Mysterium Cosmographicum,* which entails the notion of cracking the mystery or secret of the shape or pattern (or cosmography) of the cosmos. In essence Kepler found God's archetype.

The preface begins, "It is my intention, reader, to show in this little book that the most great and good Creator, in the creation of this moving universe, and the arrangement of the heavens, looked to those five regular solids, . . .and that he fitted to the nature of those solids, the number of the heavens, [and] their proportions"—a scheme easily visualized with a diagram (Fig. 6.4). The outer sphere is Saturn's, with the cube between it and Jupiter's sphere; the tetrahedron is next, followed by Mars's sphere; between it and Earth is the dodecahedron; after that the diagram is difficult to differentiate, since the planets converge quickly, but the icosahedron is between us and Venus; with lastly the octahedron between Venus and Mercury, thus completing the pattern. (Figure 5.2, which was published a later book [1619], is helpful here, since the relative distances are more easily seen. Of course this diagram came after Kepler discovered the elliptical paths of the planets, and hence embodies this fact.) Note that Kepler has some wiggle room to fit the model to the data; remember that ratios, such as 9 to 5, are only averages, since the planets still require eccentrics and small epicyclets for the model to fit the data accurately (see section 3.3). Although he speaks of solid

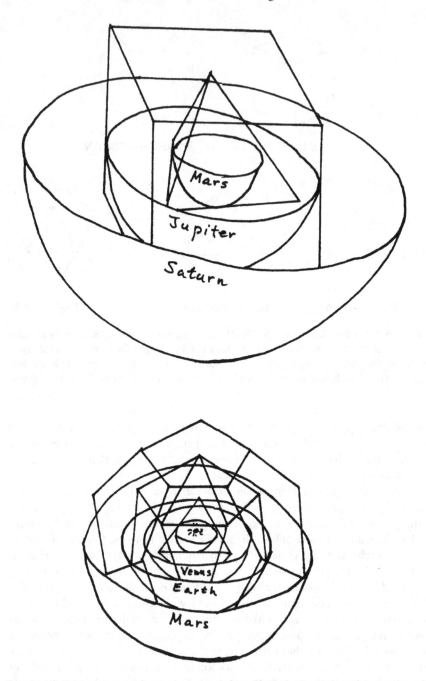

FIGURE 6.4. Kepler: the cosmic secret. A sketch from Kepler's illustration of the nesting of the five Platonic solids between the orbits of the six planets in his *Mysterium Cosmographicum* (1596). I have added the bottom diagram, since the nesting quickly converges toward the center (sun), and in the top diagram it is difficult to see the icosahedron between Earth and Venus, and the octahedron between Venus and Mercury.

spheres and solid (Platonic) bodies, it is questionable whether he is conceiving of actual physical things. More likely he was using the term simply to imply three-dimensional forms. We know that Kepler, as did Tycho, had serious doubts about the reality of crystalline spheres holding the planets in their orbits. He would eventually conceive of another mechanism, based on powers between the sun and the planets, controlling the motions of the universe, this being the first "celestial physics" (Kepler's term). All this, however, came later and perhaps at the time of the *Mysterium* he was still hedging his bets (see section 11.1).

Perhaps the most interesting aspect of the model is how it harkens back to the nesting spheres concept of Ptolemy, which, recall, Copernicus had eschewed by deducing directly the relative planetary distances (see sections 2.3 and 3.3). How ironic that Kepler reintroduces this similar conceptualization of the universe. In the end, the important empirical question is: Does the model fit the data?

Before looking at how well Kepler's scheme actually worked, let's consider some of the constraints entailed in the task. First are the given ratios of the outer and inner spheres for each of the five solids. This follows from geometry alone. For the tetrahedron it is 1 to 1/3 (or 1 to 0.333), where the outer radius is set at unity. For the cube and octahedron, the ratios are both 1 to $1/\sqrt{3}$ (or 1 to 0.577), the identity of which is certainly not intuitively obvious. Neither is the fact that the ratios for both the dodecahedron and icosahedron are also the same, namely 1 to $1/\sqrt{15 - 6\sqrt{5}}$ (or 1 to 0.795). Only three ratios, therefore, are available for fitting the five solids within the six planets, thus limiting at the start the model's ability to fit the data. The project may appear hopeless at this point, yet it did not stop Kepler.

Now, what were the data? Using Book V, Chapters 9 to 27 of Copernicus's *De Revolutionibus*, Kepler first compiled the numbers for the range of distances of each planet from the sun (namely, the aphelion and perihelion, the closest and farther points, respectively) using Earth to sun distance as unity (the astronomical unit). To convert these data to ratios that would correspond to the ratios of the solids, he used (for example), the inner number for Jupiter set to unity over the outer number for Mars, and got 0.333. Clearly this meant that the tetrahedron was placed between these two planets. Completing this task with Copernicus's data resulted in the ratios in sequence as follows:

Saturn 0.635 *Jupiter* 0.333 *Mars* 0.757 *Earth* 0.794 *Venus* 0.723 *Mercury*

This left Kepler with obvious placements, some options, and a problem.

As seen, the tetrahedron was a no-brainer: it fit between Jupiter and Mars. Also both 0.757 and 0.794 are closest to 0.795, so (from above) the dodecahedron and the icosahedron should be placed between Mars and Earth and Earth and Venus, but in which order? At some time during this exercise in fitting the model to the data, Kepler came upon this geometrical attribute of the solids: the five may be subgrouped into two classes as follows. At any vertex of the cube, tetrahedron, and dodecahedron three faces meet, whereas four faces meet for the octahedron and five for the icosahedron. Moreover, for the first group, each has a different polygon for faces (namely, a square, triangle, and pentagon); for the octahedron and icosahedron all faces are triangle. Hence, in a throwback to geocentrism, which, incidentally hearkens back to Ptolemy's rationale for placing the sun between Venus and Mars (see section 2.2), Kepler used Earth to separate these two classes of solids; thus he placed the dodecahedron between Mars and Earth (since the tetrahedron was between Jupiter and Mars) and therefore the icosahedron was between Earth and Venus.

This therefore forced the cube between Saturn and Jupiter and the octahedron Venus and Mercury. But 0.577 was not very close to 0.635 (for the former) and 0.577 is nowhere near 0.723 (for the latter). Nevertheless, these were all that remained and they had to fit; that is, they had to fit *if* this was the scheme in the mind of God in the beginning. That, however, was not in doubt: "For certainly it cannot be accidental that the proportions of the solids are so close to these [planetary] intervals"—well, at least some of them are.

As we know, the final placement (Fig. 6.4) set the sequence this way:

Saturn cube *Jupiter* tetrahedron *Mars* dodecahedron *Earth* icosahedron *Venus* octahedron *Mercury*

So, how did Kepler justify the placements of the cube and octahedron? For the fitting of the cube between Saturn and Jupiter he acknowledges the "undue discrepancy" (0.577 versus 0.635), but he dismisses the problem with this remark (which, frankly, I cannot make sense of): "However[,] at such a great distance [from the sun this] should surprise nobody." Recall Ptolemy's contention that the planetary motions get more complex near the center; Kepler seems to be arguing something like the opposite, that the order may begin to break down farther from the central sun.

Lastly, there is the fitting of the octahedron between Venus and Mercury, where he does admit that the numbers are much too far apart (0.577 vs. 0.723) to be correct. Kepler, here, makes everything fit by modifying the model. As in the other cases, the outer sphere still touches the vertices of the solid, but here he changes the placement of the inner sphere; instead of touching the faces of the octahedron it touches the sides of the square *within* the octahedron. This is seen in Figure 6.5, where I have drawn a triangle whose base is half of the square within the octahedron, and the hypotenuse is one of its faces. The height of the triangle is set as unity (the radius of the outer sphere) and the base is then the radius of this new inner sphere, which is $1/\sqrt{2}$, rather than $1/\sqrt{3}$. Thus, 1 to $1/\sqrt{2}$ is 1 to 0.707, which is a rather good fit, much closer to 0.723 than the former 0.577. Indeed, under this scheme all the solids between the planets fit the data within 5% accuracy, except for the cube. These were the hooks in the sky upon which the Creator hung the planets—in a word, God's archetype according to Kepler.

One may scoff at Kepler's naïveté with this scheme. How could the great astronomer, who went on to discover the three laws of motion found in every mechanics textbook today, take such fudging seriously? In his defense, one could make reference to more recent examples of similar manipulations of data to fit

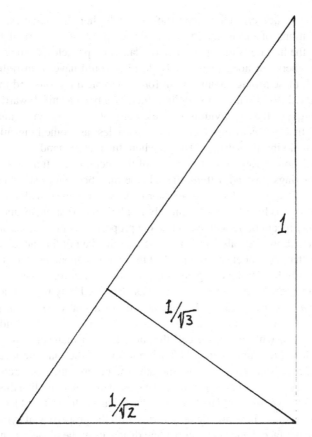

FIGURE 6.5. Kepler: fitting the octahedron between Mercury and Venus. This triangle shows how Kepler modified his construction to fit the data. The hypotenuse is a face of the octahedron, and the base is one-half the square within. The inner sphere touches the side of the square (imparting a radius of $1/\sqrt{2}$) rather than touching the face (which would give a radius of $1/\sqrt{3}$). The result is a closer fit.

preconceived schemes. An example that comes to mind is the way physicists tried to find some mathematical order to the subatomic particles. Before World War II only the electron, proton, and neutron (along with the photon) were known to exist, and from them the chemical elements were constructed. In 1930 Pauli predicted the neutrino, as discussed above. In 1935 another particle (the muon) was predicted, but not detected until after the war. The new particles (the muon, pion, and later neutrino) were initially found in cosmic rays, but with the invention of high-energy accelerators (called "atom smashers" in the vernacular) a new world of elementary particles (the kaon, lambda, sigma, et al.) proliferated over the next few decades. By the 1960s, the aggregation was dubbed the "elementary particle zoo." I was a physics student at the time, and remember the quest to find some order to this diverse aggregate of subatomic particles, with their assorted properties. A major effort involved the use of higher mathematics, especially sundry variations of group theory, as a way of providing a sort of taxonomic structure to the hundreds of supposedly fundamental particles. Some schemes fit better than others but none fit perfectly. Kepler would have been right at home playing with these arrangements—both for archetypical reasons and the fudging of data entailed. By the new millennium, there has been a shift toward two other models: the quark theory, dividing the "elementary" particles into smaller units, and string theory, picturing elementary particles as something like almost infinitely small vibrating strings. This is where the matter stands.

As a validation of Kepler's "discovery" of the secret of the universe, let's leave quarks and strings. Instead, I think the scheme may be seen to stand on its own. After all, the fit was—and still is—extraordinarily close, even with the modifications involved. It is hard to dismiss entirely Kepler's observation, as quoted above, that "certainly it cannot be accidental that the proportions of the solids are so close to these [planetary] intervals." The fit *was* close; the data fit the model, even if we know today that the model was wrong. But this was not apparent at the time. Also, given especially Kepler's theological mind set, the search for a static geometrical order as God's archetype for the universe was an exceedingly likely goal.

Mysterium was dated 1596, although it did not actually appear in print until 1597. By 1619, during more than two decades of intense and relentless work, Kepler published what today we call the three laws of planetary motion, yet he never abandoned his fundamental belief in the role of the Platonic ideal. (Notice how in Fig. 5.2, from 1619, he incorporates the ellipses into the archetype.) At some time over those 20 years he must have reasoned something like this. If I were God, I would have set the planets in circular orbits with the five Platonic solids between them, since these are the only forms with the properties of symmetry and one polygon per face; hence the number and distances of the planets would be eternally set. We know, however, that God did not assemble the universe exactly this way (bear in mind the 5% variance); thus, the Platonic scheme was only an ideal, a starting point for the final overall configuration in all its details. This leads to a deeper question: Why then did God depart from this perfect design? In a real sense, Kepler's astronomical efforts were directed to figuring this out.

6.3. Kepler and the Equant

Every student of astronomy learns Kepler's three law of planetary motion. Yet Kepler himself neither numbered them nor identified them as "laws." To him they were harmonic relations among astronomical parameters; otherwise his focus was on the larger scheme of things. The concept of laws and the 1, 2, 3 numbering scheme used today came about in the 18th century, as mechanics was being formalized in various textbooks.

The second law—that the planets sweep out equal areas in equal times—was actually discovered first, as Kepler was wrestling with Tycho Brahe's data on Mars, which he acquired upon the death of Tycho. Kepler had come to Prague in

October 1600 to work with Tycho; one year to the month later, Tycho died. In the meantime Kepler had become Tycho's favored assistant and when he died Kepler got his job as Imperial Mathematician to Rudolf II (although not Tycho's larger salary). Tycho had amassed considerable data on the positions of Mars, the most difficult planet to fit into circles (after the elusive planet Mercury). Hence late in 1601 Kepler began what he called his "war with Mars" and made a bet that he would crack the problem of the planet's orbit in a few days. It took, instead, over 3 years; around Easter 1605 he discovered the elliptical path of Mars, after having uncovered the law of areas in the process. These appeared, after much difficulty with the publication process, in 1609 in his *Astronomia Nova*—by today's standards, his greatest book. The third law was discovered in 1618 during the writing of the *Harmonice Mundi* (see next section).

The history of the finding of the first two laws is considerably complex, as historians in recent decades have themselves been wrestling with Kepler's texts. A fascinating part of that story is what I wish to tell here, involving another law of the motion of the planets using the Copernican system, but this from the point of view of the empty focus. But first we need to recall some historical background.

First is a brief review of the equant (see section 3.5). Ptolemy's effort to fit astronomical data for the planets into his geocentric system of circles upon circles (epicycles on deferents and eccentric motion) was not completely successful using only circles whose motions were centered on the centers of the circles. The model closely, but not exactly, fit the data. He could only fully fit the data by introducing a radical modification to the model: namely, that neither the center of the deferent nor the eccentric point were the center of uniform rotation; rather, another point opposite the eccentric to Earth was such a point (see Fig. 3.8). He called this point an equant, and astronomers used it, some with considerable misgivings, over the ages. Therefore, on the one hand, Ptolemy preserved the ancient Greek scheme of using circles and uniform rotation; on the other hand, by separating the previous coincidental nature of those two points, he raised the question of how far a model may be modified or stretched, so to speak, before it no longer is persuasive. Note that only the deferent requires an equant; the epicycle does not change; thus the uniform rotation of the equant is with respect to the center of the epicycle.

Copernicus surely found the equant extraordinarily unaesthetic, and one of the chief reasons for exploring alternative models of the heavens was to avoid the need for such a point. In his final construction of the heliocentric model, the centers of all the circles are points of uniform rotation. Indeed Copernicus, as a true Renaissance man, was delighted to eliminate the dreaded equant and return astronomy to its ancient pristine beauty.

Enter Kepler and his war with Mars. In struggling to fit Tycho's data to a meaningful orbital shape, he used not only circles but also ellipses as mathematical approximations of the correct shape. Ellipses were obvious mathematical shapes to use since they were classical forms, along with parabola and hyperbolas; all three had experienced a revival when ancient texts were reproduced in the

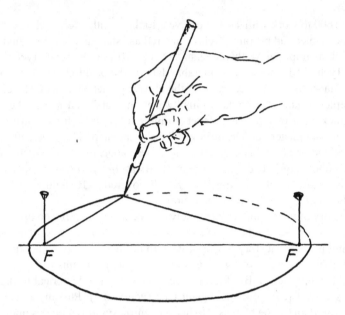

FIGURE 6.6. Ellipse construction. Kepler conceived of this handy way of constructing an ellipse, using a string tightly pulled between two pins.

Renaissance. In the course of his war, however, it did not immediately occur to Kepler to try an ellipse as *the* physical shape. One reason for this was the structure of the ellipse, which is defined as the locus of points, the sums of whose distance from two fixed points (the foci) is a constant. A simple way of drawing ellipses follows from the definition (Fig. 6.6): draw the closed path by stretching a string of given length (the constant) tied to two pins (the foci), and the resulting oval is an ellipse. This technique is found in late antique texts and was revived by Kepler and others. The elongation of the ellipse (called its eccentricity) is a function of the distance between the two foci and the constant; note that if the foci are coincidental, the path is a circle. Since an ellipse has two foci, Kepler was not sure what physical objects or parameters each focus would correspond to. Keep in mind that within his theological view of astronomy, everything had to have a purpose.

It is time to look more closely at Fig. 5.2, which has been referred to several times throughout this book. I am holding a facsimile copy of Kepler's *Harmonices Mundi*, and the diagram is isolated on a separate page (between pages 186 and 187). The pages are about 19.5 × 29 cm, with the diagram being only about 6 × 15 cm. As noted before, I believe it is the first diagram of the Copernican system that is drawn to scale. The numerical sequence of the relative distances of planets is (within the accuracy of the small drawing) drawn from the central sun to the medium of each planet. The eccentricity of the elliptical orbits is also marked by the specifications of the aphelion and perihelion of each planet. The amount of eccentricity of a planet becomes the ratio of the aphelion-perihelion range over the AU distance. Note how Venus is a thin line and that even though Saturn has the largest absolute range, Mars and Mercury each have a greater eccentricity. Kepler has not, however, drawn an actual ellipse for each planet, except for one, Mars. Looking closely, there is a dotted path from Mars's aphelion to the perihelion; really it is an eccentric circle, since given the size of the drawing he could not make an exact ellipse to scale. That he chose Mars may seen symbolic since it was the first planet for which he discovered the elliptical law; yet a more prosaic reason is clearly seen in the diagram—it was the only planet for which there was room to draw such a path! Finally, and of course, the longevity of the archetype from his first book is blatant in its specifications of the Platonic forms between the planets.

Returning now to Kepler's war, in the course of finding that the orbit of Mars actually fit an ellipse, with the sun at one focus, he also found that the other focus (what today we called the "empty" focus, since we are not burdened with Kepler's theological strictures) was an equant. Around the sun the planets speed up and slow down, moving fastest at the perihelion and slowest at the aphelion, but from the point of view of the other focus all the planets rotate with constant speed. Alas, poor Copernicus must have turned in his grave, for the equant returned to astronomy, albeit and ironically from a heliocentric point of view, as Kepler's astronomical relationships were gradually adopted. This law of the equant was often part of the elliptical law, as Kepler's laws worked their way through astronomical textbooks. It was still used as a foundational rule into the 19th century.

What ultimately happened to this "other" law? Was he right? If so, why is it not taught today? Kepler's "other" law is not exactly correct, but it is extraordinary close to being true. The calculation goes this way. Consider an ellipse with eccentricity ϵ; this is a measure of the elongation of the ellipse, where $\epsilon < 1$ for all ellipses, and $\epsilon = 0$ is a circle. Next, set up the rate of rotation of any planet from the empty focus in terms of polar coordinates around that point; doing the math results in the following equation for that rotation as a ratio (R) of two terms, the minimum speed over the maximum speed:

$$R = 1 - \epsilon^2.$$

The closer R is to unity, the closer the planet's empty focus is to being a true equant. Note that for the redundant case of a circle ($\epsilon = 0$) the motion is, as it should be, constant—a perfect equant, with $R = 1$.

The chart below is for all the planets, but remember, of course, that Kepler only knew of the first six. The first thing that stands out among these data is how amazingly close, even today, the empty foci are to being true equants. (I am sure there are many other practical "laws" of science still used today that are much less accurate than this.) In pre-Uranus astronomy, the empty focus for all the planets, except Mercury, was within 99% of being an equant. This alone must be a major reason why this "law" was considered true for so long. Even today only Pluto (along with Mercury) has an empty focus that departs from 99%, and yet both are not that far off.

Planet	ϵ	R
Mercury	0.206	0.9576
Venus	0.007	0.99995
Earth	0.067	0.9955
Mars	0.093	0.9914
Jupiter	0.048	0.9977
Saturn	0.056	0.9969
Uranus	0.047	0.9978
Neptune	0.009	0.99992
Pluto	0.250	0.9375

A very interesting point I wish to make about this fact of planetary motion—and I think this is a fascinating fact—relates back to Ptolemy. Without getting bound up in the mathematics of vector rotations and the like, let me remind you that there are ways of geometrically transforming data from the geocentric system to the heliocentric system. Keeping this in mind, here's my point: the fact that the empty focus on the heliocentric model is almost an equant vindicates Ptolemy for introducing it on the geocentric model, for this means that on both systems there really (almost) *is* an equant. Or put another way, in light of the mathematical compatibility of the two systems, Kepler's (heliocentric) discovery of the empty focus as a near equant vindicates Ptolemy's (geocentric) use of it.

And the realization of all this leads to one more question: If there really is an equant, how did Copernicus eliminate it and still make his system work? Without getting into the geometrical details, the short answer is this: technically speaking, Copernicus retained the equant in his geometrical system although it was substituted (or, maybe, better said, accounted for) by two circles (a deferent and an epicyclet, both of uniform rotation about their own centers), and so he preserved the pre-Ptolemaic ancient aesthetic principle.

6.4. Kepler's Music of the Heavens, and Beyond

What we call Kepler's third law he called a harmonic relationship between the periods and distances of the planets: namely, that the squares of their periods (T) are proportional to the cube of the average distances from the sun (D), that is, $T^2 \propto D^3$. This was discovered in March 1618 while writing the manuscript for the *Harmonice Mundi* (*The Harmony of the World*) published in 1619 (incidentally, from which came Fig. 5.2). The word *world* (*Mundi*) was still a synonym for what is later called the universe, but the key semantic matter here is about "harmony." If the word has a musical ring, you are right.

Western music has its theoretical origins in the 6th century BC, with Pythagoras's alleged discovery of a mathematical basis to musical harmony. For identical plucked strings of equal tension, the ratios of their lengths are as follows: 1/2 for the octave, 2/3 for the perfect fifth, and 3/4 for the fourth. All other ratios were deemed dissonant. Today these numbers can be correlated to frequencies. For example, on a piano the ratio of the frequency of middle C to the next C an octave higher is approximately 1/2. It is an approximation today because the original Pythagorean obsession with only the whole numbers 1, 2, 3, 4 was later modified. Such changes began in 16th-century music theory and practice, several generations before Newton. For example, Galileo's father, Vincenzio Galilei, made a major contribution to the subject; in 1581, he published *Dialogue on Ancient and Modern Music* in which he critiqued the prevailing theory of using only the ancient doctrine of number theory. One modification was to augment the system to include 5 and 6, so that the thirds and sixths would be considered consonant too. Anyone with a keyboard nearby may hear these "harmonies" by simultaneously hitting these notes: C and the next C (octave), C and G (perfect fifth), C and F (fourth), C and E (third), and C and A (sixth). Another even more radical idea, which Vincenzio proposed, was to depart from the strict confines of whole numbers. The reason was this. If, say, a piano is tuned according to the whole numbers starting with middle C, the piano will be in perfect tune only for music played in the key of C; but if a piece of music is played in any other key, it will be out of tune. And, of course one cannot re-tune an instrument every time a new piece is played in a different key. (Today, I've been told, this not a problem for advanced electronic keyboards.) This problem became increasing evident with the development of polyphonic music in the late Renaissance and Baroque eras. So Vincenzio proposed a compromise: instruments were tuned by departing slightly from the perfect whole numbers such that each key sounded nearly in tune. This method became known as tempering (see Galileo and Music, below).

Some readers may be familiar with Johann Sebastian Bach's *Well-Tempered Klavier,* written in the 18th century. The complete work consists 48 preludes and fugues; it is divided into two "books" written at different times; each book consists of 24 pieces, each in a different key (12 × 2: the 12 tones of the chromatic scale, namely all black and white notes, and the major and minor keys). Bach, too, worked on the problem of tempering, and he wrote these sets of fugues to show that his tuning system worked. His contribution to the tempering problem was the culmination of about a century and a half of debate, and formed the basis of tuning today.

Incidentally, my example of using a piano is partially anachronistic since keyboards in the 16th and 17th centuries were essentially harpsichords that pluck strings; the term *clavier* is a generic term for various varieties of keyboards. The piano, where hammers strike the strings, was invented in Bach's lifetime and he became an accomplished master of the new instrument.

An aside: Pythagoras supposedly believed his discovery of the consonant ratios 1/2, 2/3, 3/4 (i.e., octave, fifth, fourth, respectively) constituted a universal law of human nature. In modern terms we would say the perfect fifth is innate; or in today's jargon it is hardwired into our brain, or in our genes. Is this true? The argument that musical harmony is not innate is based, in part, on the variety of musical scales in music throughout the world, and hence the thesis that the Western scale is universal is seen as another example of Euro-centrist thinking.

But some musicologist point out that, despite the fact that the Western scale, as a whole, is not universal, in most musical scales throughout the world there are found some similar harmonies, such as the perfect fifth. Furthermore, recent experiments exposing infants to various harmonies show them responding positively to the perfect fifth, too. Yet, the jury is still out, and more research is needed before we affirm the innateness of Pythagoras's ratios.

Now, to link this to Kepler's idea of harmony. Within the Pythagorean cosmology was the idea that the harmonic musical ratios were implanted in the distances of celestial objects from the center. This conception became known as the "music of spheres," and its allure captivated many a scholar over the ages. The greatest to be seduced was Kepler; the essence of the *Harmonice Mundi* is the principle that there are musical ratios within various parameters of the Copernican model. After the discovery of the first two laws Kepler still passionately held to the Platonic archetype: in the *Harmonice Mundi* he remarks that "the reader should remember what I published in *The Secret of the Universe,* 22 years ago, that the number of the planets . . .was taken by the most wise Creator from the five regular solid figures." (Again: compare Figs. 5.2 and 6.4.) He notes further the very close but not exact fit of the Platonic solids to the spheres: "However, it [the model] is not definitely equal [that is, exact], as I once dared to promise for eventually perfected astronomy." Thus there must be a deeper order to the heavens—"more basic principles are needed in addition to the five regular solids"—without abandoning the Platonic ideal, "for the Creator . . . does not stray [far?] from his own archetype." The eccentricity of the orbits (now due to their elliptical forms) becomes the focus of attention in answering why God modified the original pattern. In particular, Kepler searches for ratios of variables fitting musical (harmonic) ones. After trying various possibilities, the successful program began with the discovery that the ratio of the daily speeds of Saturn at aphelion and perihelion was about 4/5— that is, the major third. Kepler went on to find (or force?) similar musical ratios among the same parameters of the other planets. For short, each planet plays a melody as it orbits the sun in its elliptical path; the greater the eccentricity of the orbit, the greater the range of notes in the tune (refer to the table with planetary eccentrics ϵ, above). Thus, for example, Venus, whose orbit is almost a perfect circle, plays one note; Mercury has the widest range (Fig. 6.7; by the way, Mercury's tune should be symmetrical as the other planets but Kepler ran out of room in the diagram). This is why God departed from using *only* circles: employing the archetype alone would result in monotonous monotonic music. This was Kepler's final discovery, the culmination of a life's work. The "secret" and the "harmony" form the bookends of his scientific life, and almost hidden within are what scientists treasure today—the three laws of planetary motion.

Despite the collapse of the bookends today, the three laws (and only these laws, although the "other" law of the equant is close) remain standing, propped up now by Newton's work on gravity. By assimilating Kepler's three laws into the inverse-square law of gravity, Newtonian mechanics ultimately formed the core of the physics of motion as it evolved into the 18th century and after.

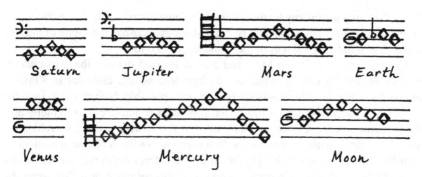

FIGURE 6.7. Kepler: music of the planets. A sketch of the tones played by each of the five planets and the moon in Kepler's conception of the music of the planets. Taken from his *Harmonices Mundi* (1619), it uses 17th century musical notation. Mercury should be symmetrical as the other planets, so it seems Kepler had to squeeze it in order to fit the diagram on the page. Note the correlation of the range of notes to the eccentricity of each planet (see Fig. 5.2).

Nevertheless, the idea of an underlying order to the arrangement of the planets was slow in disappearing. The aesthetic and abstract attraction of Kepler's scheme was not outrightly rejected. Even today, when I contemplate his idea (the Platonic ideal, minus the music), I find myself sometimes disappointed that it is not true. Figure 6.4 is a beautiful daydream, but then there is the constraint of reality.

Recall that before Kepler hit upon the geometrical scheme for the distances of the planets, he tried various numerical approaches. They all came to naught, but in the course of the endeavor he noted the relatively large gap between Mars and Jupiter, and for a time he thought another undiscovered planet might be there. With the subsequent fitting of the five Platonic solids to the cosmos, however, a seventh planet was impossible in Kepler's mind.

But others later toyed with a numerical order to the planets, the most long-lived being that conceived in 1766 by Johann Daniel Titius, professor at Wittenberg. The following table neatly summarizes the framework, process, and result:

Consider this sequence of numbers:	0	3	6	12	24	48	96
Add 4 to each term:	4	7	10	16	28	52	100
Divide by 10:	0.4	0.7	1.0	1.6	2.8	5.2	10.0
Correlate the distances of the planets:	0.36	0.72	1.0	1.5	—	5	9
The planets therefore are:	☿	♀	⊕	♂	—	♃	♄

For anyone not familiar with the symbols, the key is as follows:
☿ = Mercury, ♀ = Venus, ⊕ = Earth, ♂ = Mars, ♃ = Jupiter, and ♄ = Saturn.
Hence, there is a near-perfect fit for Mercury, Venus, Earth, and Mars, with a gap, and then a further fit for Jupiter and Saturn. This rule was picked up by another German astronomer, Johann Elert Bode, who in 1772 incorporated the idea into

his introductory astronomy textbook, and it became known as the Titius-Bode law or often just Bode's law. (It should be pointed out that, despite the fitting of the law, it is a bit of a fudge at the start of the sequence, from 0 to 3.)

The glaring gap between Mars and Jupiter implied the possible existence of another planet. Of course belief in this law demanded a disbelief in Kepler's Platonic scheme, for which only six planets were possible. Nevertheless, I see the history of Bode's law as a history of the continued commitment to Kepler's aesthetic framework for science (Kepler's spirit, if you wish)—namely, searching for a static, formal mathematical order to nature. Accordingly, some astronomers actually looked for the missing planet but were unsuccessful during the remainder of the 18th century. In the meantime, however, another planet was discovered by a German musician working as an organist in the town of Bath, England, who, in his spare time, was an amateur astronomer—this being William Herschel who, in 1781, discovered Uranus. Of course, the new planet was not between Mars and us but beyond the orbit of Saturn. As expected the query arose whether the new planet fit Bode's law. Let's do the math: $96 \times 2 = 192$; adding $4 = 196$; dividing by 10 gives 19.6. The calculation of Uranus's average distance was 19 AU—a near-perfect fit. "Kepler's spirit lives!" might be the bumper-sticker slogan today.

Not surprisingly this discovery was followed by an eager search to find the missing planet, coming to fruition on New Year's Day 1801, when Giuseppe Piazzi, at Palermo Observatory, found it (to the delight of Bode). Piazzi named it Ceres, after the goddess of Sicily. It took observations over the rest of the year to confirm its planetary nature, but early calculations gave its distance as 2.7 AU—right on target. A study of its size, however, revealed that it was extremely small. Moreover, in the following year another small planet between Mars and Jupiter was found; then two more in 1804 and 1807. For awhile, these new objects were classified as planets, so that mid-19th century celestial catalogues listed up to 18 planets. Herschel, however, called them "asteroids," estimating that they were a good deal smaller that our moon. By the late 19th century more asteroids were found, numbering in the hundreds, and hence Herschel's classification prevailed; today there are hundreds of thousands, and by estimates the total mass of them all is still much less than the mass of our moon (although, interestingly, the mass of Ceres alone is about 30% to 40% of all the asteroids).

Bode's law predicted a planet between Mars and Jupiter. The discovery of the asteroids was, at least, a confirmation of some thing or things orbiting in that space. In fact, Bode's law continued to be applied into the 19th century. A major discovery in astronomy was the prediction and confirmation of yet another planet, Neptune, beyond the orbit of Uranus. The prediction was based on anomalies in the orbit of Uranus that seemed to be due to either a modification of Newton's law of gravity at extreme distances from the sun or the possibly of a gravitational pull of something (presumably another planet) beyond Uranus. The latter hypothesis necessitated extremely complex mathematical calculations; these were performed about the same time by Urbain Le Verrier in France and John Couch Adams in England, independently. Both employed Bode's law in their calculations, revealing its longevity. Did the eventual discovery of Neptune (1846) further confirm

Bode's law? Let's see: $192 \times 2 = 384$; adding 4 gives 388; dividing by 10 gives 38.8. Neptune's distance was 30 AU, too far off. Accordingly, the demise of Bode's law began and by the early 20th century it was almost forgotten.

An interesting sidelight: not long after Neptune's confirmation, an anomaly in the orbit of Mercury evidently led to the postulation of a planet between it and the sun. After all, the anomaly in Neptune was explained by the planet Uranus, so it follow that the same should apply to Mercury's anomaly; indeed, the postulated planet was given a name (Vulcan, after the god of fire, appropriately), since some astronomers were sure it existed. There is, however, no such planet and it took Einstein's general theory of relativity to explain, by a different theory, the anomaly. The moral of this: there are no presubscribed methodological patterns in scientific discovery.

As far as I know, when the last planet, Pluto, was discovery in 1930, no one considered its possible fit to Bode's law, which is a pity, since Pluto's distance is 39.5 AU—rather close to the original prediction for Neptune. A true believer may "save" Bode's law by arguing that Pluto was, indeed, one of the original planets and Neptune was later captured. But in August 2006 the International Astronomical Union meeting in Prague proposed a new category for bodies in the solar system. Pluto was deemed to be a member of a type named dwarf planets (small spherical bodies with long eccentric orbits outside of Neptune's orbit). The body UB313 (now called Eris, first seen in 2003) was also classed as such. It has a very eccentric orbit ranging from about 38 to 97 AU, making a Bode's law calculation rather meaningless, I should think. And finally, Ceres, the largest asteroid, became a dwarf planet too. How ironic: recall that in the late-19th century Ceres had been demoted from being a planet to being an asteroid.

Galileo and Music

Galileo scholar Stillman Drake of the University of Toronto put forward the suggestive hypothesis that Galileo's exploration of experimental verification of his scientific ideas had its roots in his father's work on music theory. By criticizing the prevailing Pythagorean theory, his father was both undermining the authority of ancient texts and putting the theory of tuning to an experiential/experimental test—namely, to evaluate it by exposing it to the critical ear of the musician. An interplay between mathematics and experience, with the concomitant rejection of ancient authorities such as Aristotle, is echoed in the son's work on the physics of motion, especially the important role of experimentation.

A concrete example of the application of Galileo's musical skill to his science Drake argues is found in his experiment with falling bodies. By rolling spheres down very slightly inclined planes, Galileo was able to slow the motion enough demonstrate that the "falling" bodies were accelerating and to measure intervals of time and space, and thus deduce the mathematical law underlying this motion. Using frets from musical instruments, which, rather like rubber bands today, were not fixed, he placed the frets along the included plane so that the rolling balls made a slight click as they passed over them. With his musical sense of timing, he adjusted the frets so that the rolling balls made the clicks at equal time intervals. When he achieved this for several balls, he measured the distances between the internals and found, what he called, the rule of odd numbers. That is, setting the first interval to 1 unit, the second was 3 units, the third 5 units, and so forth. This meant that the ball was surely accelerating, and also obeying a mathematical law.

From this we get Galileo's law of falling bodies in the modern form as follows. The first time interval results in 1 unit of distance. After the second time interval the ball has traveled the 1 unit plus 3 units for a total of 4 units. After third time interval the total is 9 units. Continuing this way there are 16 distance units after 4 time intervals, and so forth. Quickly a pattern emerges: the distance units (D) traveled are equal to the time intervals (t) squared; that is, $D \propto t^2$, a fundamental law of motion today.

There is some question as to which came first, the theoretical deduction of the law or the experimental verification. Drake thinks Galileo (inductively) found it first through this experiment and then later was able to deduce it mathematically (really by geometry). Otherwise why use movable frets? If you know what you are looking for, just set it upon the included plane with the odd number rule, and prove it. But it seem to me that even if one deduces such a rule from theory, it behooves the experimenter not to force the matter, but to perform the experiment as if the rule is not known and see how, in this case, the movable frets are placed using the ear alone.

Notes and References

David Topper, "The Neutrino and the Sydney Opera House," *Leonardo* 30, No. 2 (1997), pp. 81–83.

Kepler's *Mysterium Cosmographicum* has been translated as *The Secret of the Universe*, by A.M. Duncan (New York: Abaris Books, 1981), material quoted from pp. 63–69, 105, and 157. I should point out that this is a translation of the second revised edition of 1621. See also, J.V. Field, *Kepler's Geometrical Cosmology* (Chicago: University of Chicago Press, 1988), and Eric Aiton, "Johannes Kepler and the 'Mysterium Cosmographicum,'" *Sudhoffs Archiv,* vol. 61, No. 2 (1977), pp. 173–194.

The *Harmonice Mundi* by Kepler has been translated as *The Harmony of the World,* by E.J. Aiton, A.M. Duncan, and J.V. Field (Philadelphia: American Philosophical Society, 1997), material is quoted from pp. 406–407 and 424.

On the revival of ancients texts see Paul Lawrence Rose, "Renaissance Italian Methods of Drawing the Ellipse and Related Curves," *Physis,* 12 (1970), pp. 371–404.

On Bode's law see Michael Hoskin (ed.), *The Cambridge Illustrated History of Astronomy* (Cambridge: Cambridge University Press, 1997), pp. 186–193.

For a derivation of the equation for the empty focus, see my "Kepler's Other 'Law' of Planetary Motion," *European Journal of Physics,* 12 (1991), pp. 49–50. On Copernicus's retention of the equant, see Noel M. Swerdlow, "An Essay on Thomas Kuhn's First Scientific Revolution, *The Copernican Revolution,*" *Proceedings of the American Philosophical Society,* 148, No.1 (March, 2004), pp. 64–120, appendix 2 (pp. 111–115), and Otto Neugebauer, *Astronomy and History: Selected Essays* (New York: Springer-Verlag, 1983), pp. 92–96.

On Galileo's father's music theory, see Stillman Drake, *Galileo Studies* (Ann Arbor: University of Michigan Press, 1970), Chapter 2 ("Vincenzio Galilei and Galileo"). On the included plane experiment: "The Role of Music in Galileo's Experiments," *Scientific American* 232 (June, 1975), pp. 98–104.

7
Art Illustrates Science:
Galileo, a Blemished Moon,
and a Parabola of Blood

Near the end of his life Galileo said a peculiar thing: that he wished he had been an artist rather than scientist. (Einstein said he might have been a musician had he not been a physicist.) In truth, Galileo was an accomplished artist (better at his alternative craft than Einstein at his, as far as we know); he produced lovely watercolor drawings of the moon portraying his sightings through the telescopic he built (Fig. 7.1). (I do not believe there are any recordings of Einstein playing the fiddle.) While living in Florence Galileo taught art, and one of his students was a now-celebrated artist with whom, I believe, Galileo shared one of his great discoveries. In turn, she incorporated that discovery into a painting—today her most famous one. This is the story of how that probably happened.

7.1. Galileo and Cigoli

In 1610 Galileo obtained a court position as chief mathematician and philosopher to the Grand Duke, Cosimo II, and in September he moved from his university professorship in Padua to Florence; there he became a member of the *Accademia del Disegno* (*Academy of Drawing or Design*), where he taught. The academy was a meeting place for artists, sculptors, and architects, where they conversed on matters pertaining to the arts and sciences. Important to the visual arts was geometry and its application to anatomy and linear perspective—the latter Galileo taught. At the academy he also could put forward his ideas about literature and philosophy. He wrote literary criticism and was an art critic and collector, preferring the older "classical Renaissance" style to newer movements. Scholars of Italian literature deem Galileo's *Dialogo* (*The Dialogue on the Two Chief World Systems,* one of the first scientific works written in the vernacular rather than Latin) a masterpiece. It was for Italian literature what Cervantes's *Don Quixote* was for Spanish. No doubt Galileo was an eclectic scholar of the 17th century.

The move to Florence was a step up the social hierarchy, from a lowly (and relatively low-paid) mathematics professor to a gentleman and philosopher of the Medici court. Galileo had craved such a job for years. When, in the spring of 1610, he published his initial telescopic discoveries in *Sidereus Nuncius*

FIGURE 7.1. Copy of Galileo: drawings of the moon. Galileo was a skilled draftsman, and he made several watercolor depictions of the moon as seen through his telescope.

(The Starry Message), he sought such a patronage appointment by naming his major discovery, the moons of Jupiter, after the Duke's sons: he called them the Medicean stars—and the ploy worked. Galileo's book was not only his ticket to Florence (by birth he was a Florentine on his father's side) and to court life among the nobility; it likewise propelled him into the limelight. One might say he was the first celebrity of science; of course, because of his later treatment before the Catholic Inquisition, some have also called him a martyr of science. His telescopic observations were the springboard to this fame but, in fact, Galileo was not an astronomer at heart. Before he built a series of telescopes beginning in the summer of 1609, his original work in science principally involved what we may call mathematical and experimental physics. A major breakthrough was his discovery of the law of projectile motion.

Galileo's discovery of the law of projectile motion (see details in Chapter 8) entailed a rejection of Aristotle's law that projectiles move first in a straight line from their source of motion and, after slowing down, fall vertically to the ground (Fig. 8.1). Galileo instead showed that the path was symmetrical and he performed a series of experiments to prove it. He made this discovery around 1608, when he was a professor of mathematics at the University of Padua. An important document for this story is the letter dated February 11, 1609, to a potential patron, which contains a diagram of projectile motion drawn by Galileo (Fig. 7.2). Although he did not get the patronage appointment he was seeking, the letter reveals that he knew of the symmetrical law by this time. In 19 months he was living in Florence as a member of the Medici court and teaching at the *Accademia*. (Interestingly, Galileo had once applied for a job at the academy, without success; that was in 1588, a year before he obtained his first teaching position in mathematics, at the University of Pisa.)

One of his many friends was the artist Lodovico Cigoli, who said that Galileo was his master at perspective. They had met as students of the same teacher of mathematics. About the time Galileo arrived in Florence, Cigoli was completing a major commission in Rome; between 1610 and 1612 he painted a series of frescoes in the Basilica of Santa Maria Maggiore. One painting is especially relevant here: his depiction of the *Assumption of the Virgin* on the dome of the new Pauline Chapel in the church. According to conventional iconography the Virgin's foot rests on a smooth crescent moon, this being one of her symbols. But Cigoli's depiction of the moon, in partial shadow, was like none ever seen in art. The line separating light from darkness was jagged, and the lit area presented a rough, pockmarked surface (Fig. 7.3). Why did Cigoli portray the moon this way?

In the Aristotelian cosmological framework, the moon is the dividing object between the perfect celestial world and the imperfect terrestrial world. To the naked eye it appears as moderately smooth and round yet having a slightly blemished surface—just as expected for a transitional entity lying between perfection and imperfection. But beginning in November 1609 Galileo made a series of observations with the telescopes he built. Looking first at the moon, he made the watercolor drawings depicting its cratered surface (Fig. 7.1). Another artist later copied these

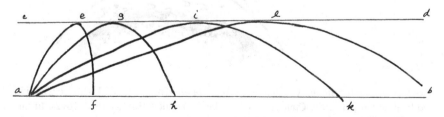

FIGURE 7.2. Galileo: drawing of parabolic projectiles. Galileo's drawing of parabolic shapes of projectiles, revealing their symmetrical forms (although the one to the far right seems to be cut off, perhaps due to a lack of space on the sheet). Detail from a letter of February 11, 1609.

FIGURE 7.3. Cigoli's moon. A sketch of the *Assumption of the Virgin* by Lodovico Cigoli in the Pauline Chapel of the Church of Santa Maria Maggiori (Rome, 1610–1612). In this painting, the Virgin's foot rests upon a Galilean moon.

Figure 7.4. Engravings of the moon: *Sidereus Nuncius*. Galileo's original watercolor sketches (see Fig. 7.1) were copied as engravings by another artist for the book.

drawings for the engravings in *Sidereus Nuncius* (Fig. 7.4). This then was the new understanding of the moon's surface, described in Galileo's book as "uneven, rough, and full of cavities and prominences." He reported seeing mountains and valleys, as well as seas and oceans, although he eventually scrapped the latter. The depiction of the moon under the Virgin's foot on the dome in the Pauline Chapel by Galileo's artist-friend Cigoli was therefore the first artistic version of what Galileo saw.

The Visual and the Sensual in Western Thought: A Primer

Galileo's extensively employed visual imagery in his first two major works, the *Sidereus Nuncius* (1610) and the *Letters on Sunspots* (1613). But this came to an abrupt end. His later publications, such as his celebrated *Dialogue* (1632) and *Discourse* (1638), rely more on written text than on visual documentation. The change was related to his career change, from a professor-mathematician (Pisa and Padua) to a gentleman-philosopher (Florence). In the exposition of his later work he tried to distance himself from the material world of visualization.

Another version of this aversion of the visual appears in post-Galilean physics, in the context of mathematical physics; often the attempt was made to derive the formulae of mechanics within a completely algebraic format, with no illustrations or diagrams. A prime example is one of the major texts in the history of mechanics—the *Analytical Mechanics* of Joseph Louis Lagrange (1788), a work containing, among other things, what today is called "Lagrange's equation." He was proud that his entire treatise contained no diagrams and few examples or applications. A later mathematician called this work a "scientific poem."

This rather odd notion is based on an idea that runs through Western thought. Its origin can be traced, at least, to Plato's idealism, which stressed mind over body. Only ideas are real; vision can play tricks on us, witness optical illusions. Indeed anything material is an illusion (recall his famous metaphor of the cave). Accordingly, the visual and mechanical arts, and their practitioners, had a low status. This may have been a factor in the dearth of experimental science from the ancient world and throughout the Middle Ages. Renaissance art theorists tried to rectify this by linking the study of perspective to the liberal arts through geometry, and, in turn, raising the social status of artists. This continued among the various academies of art that subsequently arose, but even there a remnant of Plato's dualism is found in the idea (especially popular in the French Academy of the 17th century and the English Academy of 18th century) that drawing is superior to painting. Why? Because drawing, emphasizing the line, is geometrical and relies more on the rational faculty; color, however, is more sensual and hence material, appealing to the emotions. This idea persists right into the 19th century where the famed neoclassical French artist Jean-Auguste-Dominique Ingres said that he could not adequately judge a painting without seeing the engraved (black and white) copy of it. The Impressionists, needless to say, rejected this idea with a vengeance.

We find this demoting of sensual experience even in music. For example, the French composer Hector Berlioz prided himself on being able to compose entirely in his head, not requiring the crutch of actually hearing the music—an idea that I find utterly bizarre, since music ultimately is meant to be heard. Beethoven, of course, had no choice, once he went deaf.

Galileo's rejection of the visual/sensual is a part of this story. Yet, in light of it, his statement about regretting not having chosen an artistic career seems even more peculiar.

A strange twist to this theme is found the chapter "Abstract Theories and Mechanical Models," in Pierre Duhem's otherwise masterful book on the philosophical debate about science at the turn of the last century, *The Aim and Structure of Physical Theory* (1906). Distinguishing between two approaches to science, abstraction versus visualization, he puts the dichotomy in nationalistic term. The French mind is "strong enough to be unafraid of abstraction

and generalization, but too narrow to imagine anything complex"—think of
the Lagrange tradition (mentioned above). In contrast, the English mind is
"ample but weak" and hence dependent on visuals models. Writing about a
British book on electrical theory he exclaims, "We thought we were entering
the tranquil and neatly ordered abode of reason, but we find ourselves in a
factory." The final irony of this is that Duhem was, in fact, quite an accomplished artist.

7.2. Galileo and Artemisia

In the summer of 1611, while Cigoli was working high on the scaffolding in Santa
Maria Maggiore, a visitor strolled through the church studying the frescoes. She
was a young artist, Artemisia Gentileschi, sent by her father to study several
churches in Rome. Her father, Orazio, was a successful artist working in the popular style of Caravaggio, who had a flare for dramatic depictions of biblical stories. When visiting the church, she probably saw Cigoli's depiction of Galileo's
moon. Little did she know that a chain of events that began (horrifically) in the
spring would lead to her meeting the scientist who inspired Cigoli and that she
too would depict another of his scientific discoveries in a work that posterity
would eventually deem as one of her masterpieces.

Most women artists of the Renaissance were, like Artemisia, daughters of an
artist. And she was gifted too. Her earliest extant work is a depiction of the story
of *Susanna and the Elders,* done in 1610 when she was about age 17; until fairly
recently the work was attributed to her father—it is that good. Sometime after
completing this work, Orazio hired an artist, Agostino Tassi, to give Artemisia
lesson in perspective. In May 1611 Tassi raped her. The summer trip, ostensibly
to view the frescoes, may have had a more therapeutic purpose. In March 1612
her father filed suit against Tassi with the Catholic Church. Many of the details of
the subsequent trial are known through extant Church records. What this meant,
among other things, was that Artemisia's rape became public knowledge. It is not
surprising, therefore, that not long after this ordeal we find her married off and
leaving Rome.

Early in 1613 she and her husband arrived in Florence. She subsequently
attended the academy, and in few years was elected a member—one of the few
women ever accepted. Recall that Galileo had also recently moved to Florence to
work for the Duke, and in October 1613 he became a member of the academy,
teaching perspective. He probably was her teacher. We know they met, for among
her 28 extant letters is one written to Galileo many years later.

She also worked for the Duke. In particular he commissioned a painting of the
biblical story of *Judith and Holofernes.* When she arrived in Florence she had with
her a canvas on this theme on which she had started working sometime in 1612;
she completed the painting during her first year in Florence. The story of the heroic

Judith saving her people by seducing the enemy general Holofernes in his tent, getting him drunk, and then cutting off his head, is interpreted (not surprisingly) by some historians as an act of psychological retribution for the rape (Fig. 7.5). True or not, surely the way she depicts Judith hacking off the general's head

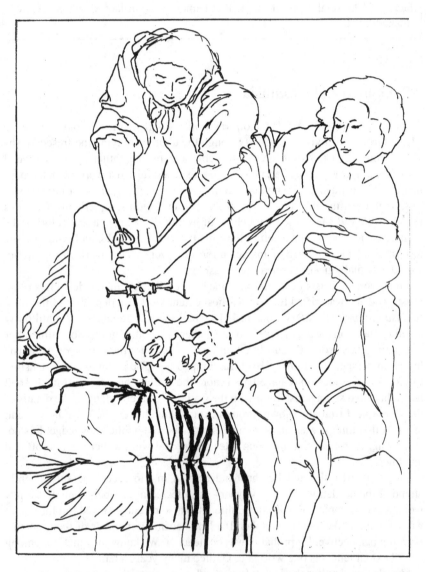

FIGURE 7.5. Artemisia: *Judith Beheading Holofernes* (first version). Oil on canvas, $62\frac{3}{10}$ by $49\frac{1}{5}$ inches (c. 1612–1613), Museo Nazionale di Capodimonte, Naples. In this work the blood of Holofernes oozes out, staining the bedsheets.

was grotesque and powerful, quite unlike the way most male artists portrayed her. Usually Judith was shown as a squeamishly meek young thing disgusted with the job she had to do and cringing at the sight of blood. This, for example, was the way her father's mentor, Caravaggio, portrayed her in a famous work (*Judith Beheading Holofernes*, 1598–9). But Artemisia shows Judith going at it with gusto; even her maidservant joins in the slaughter, rather than just standing back (as in Caravaggio's work) and literally holding the bag (for Holofernes's head).

Obviously, Cosimo in due time saw her painting and for whatever reason like it and commissioned a copy in 1620, which she completed the following year just before the Duke's death (Fig. 7.6). She was duly paid for the work, but only with the help of Galileo. We know this from that one letter to him, dated October 9, 1635, by which time he was living under house arrest following his trial before the Inquisition. In it she is requesting his help in getting paid for a work recently sent to Cosimo's successor, Ferdinand II. The reason for the appeal to Galileo: she reminds him that once before she "obtained an excellent remuneration" for "the painting of that Judith which I gave to His Serene Highness the Grand Duke Cosimo" because of Galileo's help. Clearly she is referring to the second *Judith and Holofernes* (1620–1), for which she informs us that Galileo's intercession resulted in her getting paid. The importance of this letter is that it reveals that Galileo knew of Artemisia's painting of *Judith*. Probably he saw both versions. (What forays Galileo made on her behalf in 1635, whether she ever got paid, and even what this later unpaid painting was, we do not know.)

A comparison between the 1612–3 and the 1620–1 versions of *Judith* (Figs. 7.5 and 7.6) reveals several differences, the most striking surely being the introduction of the splattering blood in the latter work. Let us look closely at the blood, less from a visceral and more from a geometrical point of view. Note how the streams form a series of arcs, mostly symmetrical and quite similar to the projectiles drawn by Galileo (Fig. 7.2). If we trace these arcs as geometrical abstractions, they appear as in Figure 7.7. In transferring these arcs to graph paper and identifying individual (x,y) pairs, I found that they closely fit equations for parabolas (namely, $y = Qx^2$, where the constant Q is a function of the particular shape of the arc of the given parabola). The one shape that does depart most from a perfectly symmetrical parabola, interestingly enough, is similar to the one asymmetrical arc drawn by Galileo (Fig. 7.2), which may have been due simply to a lack of space on the page. Otherwise, the fits, I am convinced, are just too close to be entirely accidental. This gushing blood I thus believe discloses yet another link, a quite fascinatingly subtle and pictorial one, between Artemisia and Galileo.

The second *Judith* was executed after Artemisia met Galileo. We know that he discovered the parabolic shape of a projectile about 1608. But he did not publish it until 1638, when it appeared in *Two New Sciences*. He writes of his discovery, his first published presentation of this important law, this way: "It has been observed that missiles or projectiles trace out a line somewhat curved, but no one has brought out that this is a parabola. That it is . . .will be demonstrated by me." Galileo probably penned that sentence about 1630 or 1631, in

FIGURE 7.6. Artemisia: *Judith Beheading Holofernes* (second version). Oil on Canvas, 78 by 63 inches (c.1620–1621), Uffizi, Florence. In this work, the artist moves further away from the subjects of the first version, and adds the gushing blood spraying onto Judith's arms and breasts.

the first draft of the manuscript of his book. Then in 1632, Bonaventura Cavalieri, a mathematician who studied under one of Galileo's best pupils, published a book on parabolas in which he put forward Galileo's projectile law, complete with a diagram of the symmetrical arc. Galileo was extremely upset; after all, the law was still in his manuscript. But he was placated when learning that Cavalieri mistakenly thought that Galileo had already announced the law. This

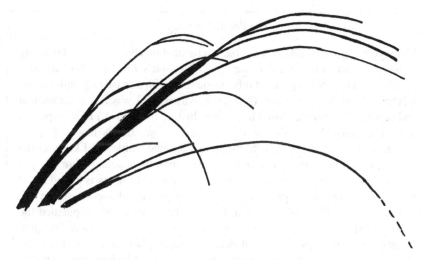

FIGURE 7.7. Parabolas of blood: *Judith and Holofernes.* Tracings of the paths of the spurting blood from Artemisia's second version of *Judith.*

incident shows that although Galileo was saving his law for his book, he was not hiding his discovery from his students. We may, accordingly, include Artemisia in this group. Moreover, although Cavalieri's diagram has been called the first published illustration of Galileo's discovery—and it surely is the first *scientific* illustration—yet, we may add the caveat that it is not the first depiction of the parabolic arc.

Hence we arrive at the following scenario. In her second version of the *Judith,* executed after she met Galileo and after he shared his projectile law with her, Artemisia pictorially recorded this law in her painting by depicting the spewing blood as parabolic arcs. Perhaps her motivation was to achieve a heightened realism, a goal synonymous with the Caravaggio school of art. Like Cigoli in his depiction of Galileo's moon, Artemisia was likewise informed by a discovery of Galileo, and she presented the first artistic rendering of it—indeed, the first "published"/public visualization. Cigoli's art came after the publication of the *Sidereus Nuncius.* Artemisia's "illustration" was made about midway between the discovery and Galileo's publication of the law of projectiles.

Unfortunately, none of this sheds light on questions that must remain forever unanswered: What did Galileo think of Artemisia's *Judith,* and especially what was his view of the spurting blood? We do know something about his view of the art of the time. It comes as no surprise to hear that he preferred art exhibiting order, simplicity, and balance and objected to various distortions favored among the more experimental artists (the so-called "Mannerists") of his time. In this Galileo shows himself to be a classicist, which perhaps reflects his love of geometry. It certainly tells us something about what sort of artist he would have become had he pursued that career instead.

The Parallel Fallacy

Many years ago after taking a course in classical mechanics, I was browsing in the university library and came across Newton's *Principia* (in English). Pulling it off the shelf I perused it for some time with a sense of befuddlement mixed with incomprehension. (Note: this was when I was a science student and before I had been exposed to any real history of science.) I had expected to find, at least, Newton's law of motion, $F = ma$, or his law of gravity, $F = GmM/r^2$, but neither was transparently spotted. Little did I know at the time that Newton's laws of motion as we write them today were not put forward until the mid-18th century. And that the law of gravity is buried in the text, and hence not expressed in the *Principia* as above (see Newton, Cavendish, and Newton's Laws, in Chapter 10). Later, when I pursued the history of the subject, I realized that there is a disjunction between original scientific writings and the distillation that takes place as textbook writers abstract and synthesize what scientists have wrought. Much of this is a theme of this book. It also offers some insight into what I believe is a fallacy often put forward about the differences between art and science.

My argument has it origin in statements made about art and science that are of this form: An artwork Y would not exist without the artist X, whereas a scientific theory Q would (eventually) exist without the scientist P. Some concrete examples:

X = Leonardo   Y = *Mona Lisa*

X = Handel   Y = *Messiah*

P = Darwin   Q = theory of evolution

P = Einstein   Q = theory of relativity

The fallacy is based on the truism that without the individual artists and musicians no one would have painted the *Mona Lisa* or written the *Messiah* whereas eventually someone would have discovered the laws of evolution or relativity or mechanics. The fundamental problem with this line of reasoning, however, is this: the entities being compared are not the same category of things in the arts and sciences. To make this concrete: the physical painting *Mona Lisa* and the musical score *Messiah* should instead be compare to the book *Origin of Species* and the scientific article "On the Electrodynamics of Moving Bodies," and other articles by Einstein. If Newton had not lived (indeed, he almost died, having been born prematurely), the *Principia* surely would not exist. In this way comparisons are among what I called artifacts, the actual works of the creators, and these are unique to each of them. Although scientific theories, of course, have their origins in their artifacts, it is a fallacy to compare the theories of science with the artifacts of art, the result being what I call the parallel fallacy.

This deduction raises further questions, such as this: Is there a category in the artistic realm that corresponded to that of "theory" in science? There are "theory" books in the arts, such as sketchbooks, workbooks, how-to-draw books in the visual arts, as well as music books on harmony, and so forth. So I do see a parallel here between the arts and sciences, but I would not wish to push this too far. For there is a propensity, once the similarities between them are broached, to overstate the similitude. Certainly there is not the whole realm of empiricism and experimentation in most of what constitutes the arts. Also, much of the beauty of music can be perceived on the aural plane alone, with little or no knowledge of pitch, rhythm, sonority, musical texture, or form, let alone the underlying chord progressions, or even especially the mathematical underpinnings of these harmonies. With science, however, there is no aural plane for the novice to comprehend; science cannot be appreciated as music and art can. Historically put, notwithstanding Kepler's effort (see section 6.4 and Fig. 6.7), there is no music of the heavens, and, despite Newton (see section 9.3), no music of the spectrum of colors.

Notes and References

I first told this story of Galileo and Artemisia in a paper written with a student: David Topper and Cynthia Gillis, "Trajectories of Blood: Artemisia Gentileschi and Galileo's Parabolic Path," *Women's Art Journal* 17, No. 1 (Summer, 1996), pp. 10–13.

Artemisia's letter to Galileo is reprinted in the appendices to Mary D. Garrard, *Artemisia Gentileschi: The Image of the Female Hero in Italian Baroque Art* (Princeton, NJ: Princeton University Press, 1989), which also contains the transcript of the rape trial. The most thorough study of the trial is Elizabeth S. Cohen, "The Trails of Artemisia Gentileschi: A Rape as History," *Sixteenth Century Journal* 31, No. 1 (2000), pp. 47–75. On Galileo and Cigoli, see Miles Chappell, "Cigoli, Galileo, and *Invidia*," *Art Bulletin* (March, 1975), pp. 91–98. Cavalieri's illustration of the parabolic trajectory is reprinted in William B. Ashworth, Jr., "Iconography of a New Physics," *History and Technology* 4 (1987), pp. 267–297, on p. 274.

On Galileo's art criticism, see Erwin Panofsky, *Galileo as a Critic of the Arts* (The Hague: Martinus Nijhoff, 1954). On Galileo and visual imagery, see Mary G. Winkler and Albert Van Helden, "Representing the Heavens: Galileo and Visual Astronomy," *Isis* 83 (June, 1992), pp. 195–217. For more on Galileo and artists, see Eileen Reeves, *Painting the Heavens: Art and Science in the Age of Galileo* (Princeton, NJ: Princeton University Press, 1997).

More on my parallel fallacy thesis is in David Topper, "The Parallel Fallacy: On Comparing Art and Science," *British Journal of Aesthetics*, 30, No. 4 (October, 1990), pp. 311–318. After publishing this argument, I was pleased to find a parallel thesis by Owen Gingerich, "Circumventing Newton: A Study in Scientific Creativity," *American Journal of Physics*, 46, No. 3 (March, 1978), pp. 202–206, reprinted in his book, *The Eye of Heaven: Ptolemy, Copernicus, Kepler* (New York: American Institute of Physics, 1993), Chapter 25.

Pierre Duhem, *The Aim and Structure of Physical Theory*, trans. Philip P. Wiener (New York: Atheneum, 1962); this is a reprint of the second edition of 1916. On Duhem's art, see Stanley L. Jaki, *The Physicist as Artist: The Landscapes of Pierre Duhem* (Edinburgh: Scottish Academic Press, 1988); I reviewed this book in *Leonardo* 23 (1990), p. 454.

The Berlioz anecdote is from Charles Rosen, "On Playing the Piano," *The New York Review of Books* (October 21, 1999), pp. 49–54.

8
Ensnared in Circles: Galileo and the Law of Projectile Motion

Galileo is credited—and rightly so—with a momentous discovery in the physics of motion. A projectile, thrown by hand or shot from a cannon, moves along a symmetrical path that traces the arc of a parabola. Such perfect mathematical motion takes place only in an ideal world with no friction, yet a projectile approximates this in the real world. This discovery was an important component in the development of what in time became—to use the vernacular—rocket science.

What is not widely known, is this: Galileo did not truly believe in the parabolic path; instead, he thought it was only an approximation to the true geometrical form. Moreover, it was embedded in a theory that asserted that gravity is an illusion.

8.1. The Problem with Projectiles

The motion of projectiles, like most everything else in what constitutes ancient physics of motion, had its origin in the writings of Aristotle. All motion near Earth was categorized either as natural (up and down only, namely, gravity and levity) or unnatural (requiring a continuous and contingent push). An understanding of a projectile's path, Aristotle correctly realized, requires that its motion be divided into two components; erroneously, however, he thought that these two components could not act at the *same* time, believing that the stronger one would always overpower the weaker one. As a consequence, he described a projectile as initially moving along a straight line from its source of motion until it slows toward a stop; only then, as this power weakens, will downward natural motion prevail, so that the projectile then falls vertically to the ground. Aristotle's writings are not always clear on the cause of the initial straight-line motion, but it seems that he thought this unnatural motion was propelled by whirlpools of air circling around the object, giving it the required contingent and continuous push. This theory of projectile motion prevailed from the late ancient world through the Renaissance with little change and only a few challenges.

How projectiles move was not merely an intellectual exercise for the 17th century. It had a practical application in warfare and the development of cannons. The key problem was to know the correct angle to aim a cannon in order to hit its

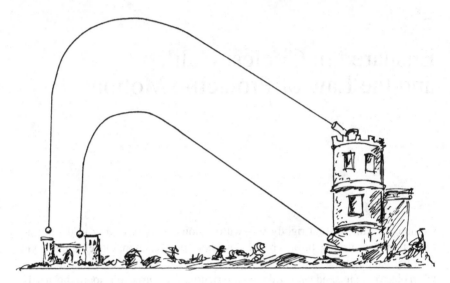

FIGURE 8.1. Projectile motion: 17th-century diagram. A sketch of the illustration of projectile motion from a 17th-century military textbook, indicating the longevity of Aristotle's concept of motion.

target. Diagrams illustrating Aristotle's two-part motion, some found in military manuals, appear well into Galileo's time (Fig. 8.1). Galileo himself was aware of such applications; when he taught at the universities of Pisa and Padua, many of his students were young noblemen pursuing a military career. He said that he learned much of practical value from conversing with the artisans and engineers who worked at the Arsenal in Venice. The University of Padua was essentially the university of the Venetian Republic, and Venice was only about a half-day's journey from Padua. The Arsenal was a military workshop and naval dockyard that one historian has characterized as "an industrial assembly line for ships, sails, and weaponry centuries before Henry Ford cobbled together his first Model T."

Galileo's teaching duties, however, did not explicitly include Aristotelian physics. This may seem strange today, but Galileo taught in the faculty of mathematics, which was separate from philosophy, and the latter was the domain of (Aristotelian) physics. Nevertheless, when Galileo taught the mathematics of the heavens (that is, astronomy) the physical nature of the cosmological model must have been broached, and this would require some discussion of terrestrial motion. We also know that to supplement his income (mathematicians, incidentally, were at the bottom of the university hierarchy), he tutored a number of students, and he exposed them to his new ideas and experiments on the physics of motion. These new ideas departed radically from what the Aristotelians were teaching.

An Experiment in Perception

Although not specifically applicable to the subject of this chapter, I find the following exceedingly interesting for reasons that will be immediately clear.

In human visual perception there is a phenomenon called "binocular rivalry." It is confirmed by the following experiments. In the first case, a different picture is shown at the same time to each eye of a subject; the result is that the subject sees only one of the pictures. For example, if one eye is shown horizontal lines and the other vertical lines, the subject sees only either the vertical or horizontal lines, not a grid. Thus the two images are not combined; only one "wins." In the second case, if one eye is shown a steady image and the other a different flashing image, the subject sees only the flashing one; apparently the novelty of the flashing image dominates the perception. Again, there is no combination of the images in the actual perception—one takes over in human visual experience.

8.2. Galileo's Abstraction

Galileo's innovation in mechanics begins with an abstraction. In order ultimately to derive a mathematical foundation for motion in the real world of earth, water, and air, Galileo abstracted a pristine world of simple objects moving along geometrical shapes in a void. He was convinced that only in this ideal world could the laws of nature come to light, otherwise they got gummed up with real-world friction and the like.

Two more intellectual breakthroughs led to the first correct law of projectiles. Kepler had coined the term *inertia* (Latin for inert or lazy) for the fact that heavy objects require a push to get them moving; but he also thought, as Aristotle had, that a force or power was thus required to keep an object in motion. Galileo, conceptualizing within an abstract world devoid of all resisting media (such as friction), reasoned that an object, once given an initial push, would move indefinitely until and unless an external force stopped it. The introduction of a resisting medium only slows the object to a stop. Inertia thus entails the propensity of a body *both*-to stay at rest and (importantly) to stay in motion. Here it is in Galileo's own words (although I should point out that the context is rotational motion, the importance of which will loom large later): in a vacuum, "a heavy body ... will maintain itself in that state in which it has once been placed; that is, if placed in a state of rest, it will conserve that [state]; and if placed in movement ... it will maintain itself in that movement." Sounding modern, this is a lucid definition of inertia seemingly right out of a textbook today. (Although Galileo never actually used the term "inertia" he did express the concept, and I henceforth will call it such.)

Galileo corrected Aristotle's other error: he persuaded himself that more than one power may act on a moving body at the *same* time. For example, if an object is pulled simultaneously by two powers or forces in two different directions 90° apart, it will move between them, along a diagonal path. "Such motions," he wrote, "in mixing together do not alter, disturb, or impede one another." Thus Galileo realized that a projectile is moved by two simultaneous powers: inertia and gravity. Consider a projectile launched horizontally: there is a horizontal (inertial) component of motion, moving the object along at a constant rate, and a vertical (gravitational) component, accelerating the object downward. That the latter motion is accelerated he had also deduced from the abstract world of motion in a void, for without a resisting medium, a falling body would continually change speed and hence (by definition) accelerate. The combination of these two powers acting simultaneously on the body results in the path of the projectile being the arc of a parabola—one of the fundamental mathematical curves from ancient geometry.

In the early 1970s, Stillman Drake discovered in Galileo's manuscripts at the National Central Library of Florence that Galileo did more than logically deduce the law of projectile motion. He actually performed a series of delicate experiments in which he measured the paths of projectiles in his laboratory. This hands-on approach was consistent with Galileo's practical bent, as seen in his acquaintance with the artisan-engineers at the Arsenal in Venice.

Figure 8.2 is from a key page of these manuscripts, revealing Galileo's measurements of the parabolic paths. Drake set the date of this page at 1608, possibly in the summer. This was much earlier than previous scholars had proposed for this discovery, which Galileo did not publish for another 30 years.

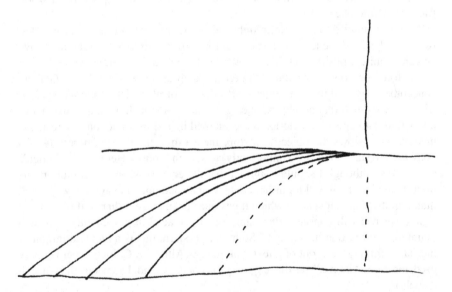

FIGURE 8.2. Galileo: manuscript on projectile experiments. A detail from an important page (c.1608) from Galileo's notebooks, revealing his experiments on projectile motion, and showing their parabolic paths.

Further evidence for the earlier date of Galileo's manuscript comes in the form of the diagram drawn by Galileo (see Fig. 7.2) in the letter to a potential patron dated February 11, 1609. This diagram, depicting projectiles launched at various angles, more clearly reveals the symmetrical paths of the projectiles. An important consequence of Galileo's discovery is that a projectile launched at a 45° angle produces the maximum distance. The military application of this to cannonballs was not to be lost on a possible patron; however, the letter did not have the desired result, since Galileo did not get the job.

Much later, in 1638, the law appeared in *Two New Sciences,* Galileo's last and arguably his major work, published in Holland from a manuscript smuggled out of Italy when he was under perpetual house arrest following his trial before the Inquisition. Of his discovery he writes: "It has been observed that missiles or projectiles trace out a line somewhat curved, but no one has brought out that this is a parabola. That it is . . .will be demonstrated by me." And so, about 30 years after discovering this important law, Galileo formally (and finally) published the correct path of projectiles, advancing scientific knowledge and technical application. (On this law and the artist Artemisia Gentileschi, see the previous chapter.)

Deriving the Parabola

The following is a thoroughly modern derivation of the law of the parabolic projectile. Although Galileo worked exclusively within a geometrical framework, his deduction is easily derived in modern algebraic notation. The horizontal component of a horizontally launched projectile moves over equal distances (x) in equal times (t), due to inertia; thus its constant speed v is, $v = x/t$. The vertical ("falling") component, moving y distances, is accelerating, so that y is proportional to t^2. (See essay on Galileo and music, Chapter 6. This law of falling bodies Galileo discovered along with the fact that objects fall independently of their weight.) It thus follows that y is proportional to x^2, clearly a parabolic function.

Scholars poring over Galileo's manuscripts have not yet reached a consensus on exactly how he arrived at it.

8.3. The Relativity of Motion and the Rotating Earth

From his discovery of the parabolic path of a projectile Galileo made one of the most important deductions in the science of motion, which became the core of the principle of relativity. He applied this to his defense of Copernicus, where he was bound to explain how our everyday earthly experience takes place on an Earth rapidly moving around the sun.

One of the key arguments against a moving Earth (beyond the lack of stellar parallax; see section 2.3), an argument also found in antiquity, is based on falling bodies. A stone dropped from a tower should not fall directly to the bottom if Earth moves; rather, it should fall somewhere "behind" the tower, so the logic goes. But Galileo countered this with the following thought experiment. He presented it in the context of motion on a moving ship. (Einstein would later use trains and elevators.) Galileo argued that a stone at the top of the mast of a moving ship not only would be compelled to move downward (by gravity) but— and this is the key insight—there also would be a component of inertia in the direction of motion of the ship, since both the ship and the stone share this motion; when dropped, this horizontal component would act on the stone just as if it were a thrown projectile. This horizontal component, along with gravity acting vertically, *simultaneously* moves the stone along a projectile path such that the stone hits the deck of the ship at the spot directly below the mast, as the moving ship catches up with it. From the point of view of an observer on the mast, the stone just falls vertically to the bottom (Fig. 8.3). So, extrapolating from the ship to the moving Earth, despite our moving through space, objects dropped on Earth appear to fall straight down. As noted, Galileo's insight contains the foundation of the modern principle of the relativity of motion.

A modern stroboscopic photograph beautifully illustrates Galileo's insight (Fig. 8.4), by sort of dissecting the temporal sequence of motion. Two balls are simultaneously ejected: ball A is dropped and ball B is a projectile shot horizontally. Note that although ball B is moving along a parabolic path, its vertical component is following the same law of fall as ball A (namely, the distance being proportional to the time squared) and independently of its horizontal (inertial) motion. This also shows the principle of relativity, since the same path would be followed if ball B were dropped from a moving source, and hence from the point of view of that source, the ball would appear merely to fall straight down.

An aside: some of the most reproduced stroboscopic pictures of motion found in science textbooks were taken by Berenice Abbott, a 20th century photographer famous for her cityscapes, especially of Paris and New York, but who also was

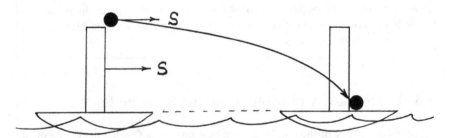

FIGURE 8.3. Galileo: relativity of motion. This thought experiment illustrates the relativity of motion. As the boat moves with uniform speed *S* through the water, the ball, whether remaining at the top of mast or falling, also moves horizontally (by inertia) with speed *S*. If falling, it follows a path from left to right, from the top of the mast, reaching the deck as the boat moves with the same speed *S*. From the point of view of someone on the boat, the ball merely falls straight down.

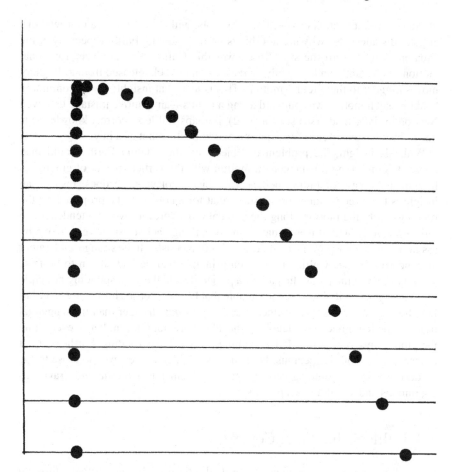

FIGURE 8.4. Stroboscope photograph of motion. A sketch of a stroboscope photograph illustrating the independence of the horizontal and vertical components of projectile motion. Note that the rate of fall (acceleration) is the same for both balls.

greatly influenced by science. Mainly in the 1940s and 1950s she produced memorable photographs demonstrating various physical laws of nature.

Having provided a justification for a moving Earth, Galileo was left with a problem: if Earth really rotates once a day, why are we not flung off it? To answer this correctly requires an understanding of what later is called centrifugal force (see section 11.2), and for this a clear concept of linear inertia is imperative. What is extraordinarily interesting about Galileo's reasoning on this question is the mélange of insight and error.

He begins with the case of a bottle of water tied to a cord and being spun in a circle. If there is a hole in the bottle, the water will "spurt forth." The same sort of motion, he points out, is involved in slingshots, with which boys can throw rocks at great distances. So why are not things on the spinning Earth, especially at the equator, flung toward the sky? To answer this Galileo first analyzes rotational motion: importantly and correctly he reasons that a rock released from a slingshot moves *tangent* to the circular motion. This is a critical insight; it was commonly held that such motion was outward along a radius—an intuitive mistake that even Newton initially made (see section 11.2). Perhaps Galileo's correct knowledge is due to his own experience (or experiments) with slingshot as a boy.

With his insight, the problem of objects on the rotating Earth should then reduce to combining this tangential motion with the vertical power of gravity, as he did in the case of falling bodies. But this is not what he does. Instead, an insight is followed by an error: he states that for an object to be projected off the rotating Earth, the motion along the tangent must "prevail over the tendency" to fall. Amazingly, with this affirmation, he is harking back to Aristotle's theory, by assuming that two powers cannot act simultaneously. It is strange and rather disconcerting to see Galileo revert to a principle that he had shown to be false when he derived the parabolic path of a projectile and the corresponding principle of relativity. I find it almost unfathomable that he does not see his error. Nevertheless, he goes on to assert that since gravity is *always* stronger than the tangential motion (the legitimacy and details of this claim are not relevant here), everything therefore stays on Earth. Hence nothing flies off the rotating Earth—but, of course, the "proof" is specious. Nonetheless, Galileo comes ever so close to the modern theory of deducing centrifugal force from linear inertia and gravity as continuous and simultaneous powers.

8.4. Galileo's Inertia in Context

Close, however, only has meaning as Galileo's insights are filtered through Newton's physics. In Galileo's mind his discoveries led to an explanation of celestial motion and gravity, at once. Here's how.

The previous quotation on inertia, recall, was about circular motion. A concrete example of Galileo's realization of inertia in action is a delightful experiment anyone can perform (see section 4.4), which appears in *Dialogue on the Two Chief World Systems* (1632). Float a ball in the center of a bowl of water and carefully rotate the bowl. One finds that the ball's orientation with respect to the room remains fixed despite the rotational motion of the water and bowl in which it floats; this neatly illustrates inertia with respect to rotation, since the ball remains at rest. Concomitantly, an object in circular motion will retain that motion unless it is externally suppressed. Certainly the concept of inertia was central to Galileo's maturing scientific thought, but the question arises as to the nature of this concept. It seems that he did not conceive of it as today's textbooks do—namely, as linear motion. This is a matter of some debate among historians but I am convinced that the consensus is right: Galileo's conceived of inertia only as a circular phenomenon.

The framework of Galileo's work in mathematical physics was the Copernican system, specifically the simplified heliocentric model. Here the sun sits at the center of the sphere of the fixed stars, and the planets, which now include Earth, move in circles around this center. As a first approximation—for example, accounting for the yearly motion of Earth or the 12-year period of Jupiter—the model is sufficient; for a complete explanation, however, more circles (eccentrics and epicyclets) are needed because of numerous other minor motions (see section 3.5). Just as Galileo abstracted and simplified the physics of motion to arrive at his mathematical laws, so he condensed and reduced cosmology to circles centered on the sun, thus ignoring Kepler's discovery of elliptical planetary orbits. Combining this abstraction with that of inertia as a *circular* phenomenon results in a celestial physics as follows: the planets orbit the sun by inertia alone, not requiring, therefore, any other powers such as Kepler had proposed between the sun and the planets or even the planets and their moons (see section 11.1). All traces of occultism (action-at-a-distance) Galileo thus banished from the cosmos. In this framework, only gravity, as a *local* power of unknown cause, remained.

But there's more: Galileo even found a way of eliminating gravity completely. Consider the following fascinating argument on the relativity of motion applied to the entire rotating Earth. It appears as a geometrical proof in the *Dialogue*. In Figure 8.5, A is the center of Earth, BI is on the surface of Earth (therefore AB is Earth's radius), and CB is a tower. As Earth rotates, the top of the tower moves through points F, G, H, etc. Now if a stone is released from the top the tower, it falls on the moving (rotating) Earth along the path of an arc from C to I, terminating at the surface of Earth; assuming a transparent Earth, ideally the stone would proceed to the center of Earth, point A, making a half-circle centered on point E. Although in the end this argument is wrong, it does contain the clever artifice of geometrically conceiving of a transparent Earth, something that Robert Hooke later utilized in his significant contribution to a celestial physics, which influenced Newton (see section 10.4).

However wrong, Galileo's deduction from this geometrical demonstration is worth looking into, since it is a window deep into his frame of mind on the physics of motion. He proves that the distances between arcs C, F, G, through D, centered on the semi-circle A, are equal to the arcs from C to I centered on a semi-circle at E. (I leave it to the reader to confirm his proof; it's a neat problem in geometry.) From this he infers that there is no difference between the stone stay-ing on the tower (moving from C to D as Earth turns) and the stone falling to Earth (from C to I); in both cases they move along circles through equal arcs. In his words, "Whether remaining on the tower or falling," the stone "moves always in the same manner; that is, circularly, with the same rapidity, and with the same uniformity." He speaks of his proof as a series of "curiosities." As well, he calls it a "marvel," for this means that the straight-line (vertical) power of gravity is eliminated, being really a deception due to the motion of Earth. Indeed "the true and real motion of the stone is never accelerated [namely, falling] at all, but is always equable and uniform" in its circular motion. Accordingly, everything is subsumed under circular inertia, and hence "straight line motion [both vertical and horizontal] goes entirely out the window and nature never makes any use of it

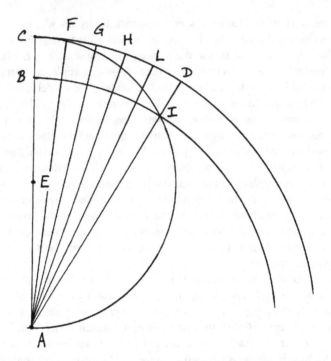

FIGURE 8.5. Galileo: the illusion of gravity. Geometrical diagram for explaining Galileo's (ultimately erroneous) argument that gravity is an illusion. Point *A* is the center of Earth and therefore the surface of Earth is the circle *B* to *I*, etc. The falling stone starts from point *C*, and as the stone falls from *C* to *I* on a rotating Earth, it is really moving in a circle toward the center of Earth (*A*). All motion is circular, and hence gravity (as straight-line motion toward Earth) is an illusion.

at all." All that remains is natural (rotational or circular) motion, since the stone "really moves in nothing other than a simple circular motion, just as when it rested on the tower it moved with a simple circular motion."

This is a remarkably ingenious and attractive model. It is easy to see why Galileo found it so appealing and convincing. If true, it would be a "marvel": it beautifully unites a terrestrial physics with a celestial motion on the Copernican system, thus eliminating the dualism inherent in the vertical motion of gravity and the circular motions of moons and planets. Or, said another way, gravity is an illusion. Too bad it is wrong. (Just as it is too bad that Kepler's "secret of the universe" is wrong; see Chapter 6.)

So what is wrong with the proof? The ultimate fault is this: rotational motion itself really is accelerated motion. But this was not fully comprehended until the later conceptualization of linear inertia among Descartes, Huygens, Hooke, Newton, and others, which further supports the view that Galileo only conceived of circular inertia.

Independently of this retrospective error, however, there are problems within Galileo's own framework. First, he is not being consistent in the way he applies circular inertia. Consider the case of moons orbiting planets or planets orbiting the sun. It is true that from the point of view of circular inertia, these circular motions entail no explicit occult powers between the orbiting objects and those at the centers; nevertheless, there is an implicit connection in that the orbiting objects move in their particular circles around the central bodies, rather than in any other circle about any other geometrical center (and there are an infinite number of possibilities). Put simply, our moon orbits Earth rather than moving in some other circle because Earth is there! And this brings me back to Figure 8.5 and point E. Although the falling stone is moving toward the center of Earth (point A, which is a physical, namely gravitational, center) the stone is "orbiting" point E, which is only a geometrical point, with no physical meaning. Why orbit around E? This seems to contradict the way circular inertia is applied to planets and moons.

Second, there is Galileo's omission of the parabolic projectile path. Should not the path of the falling stone from C to I be identified as a parabolic arc? As noted above, there is a strange 30-year gap between his discovery of the law (about 1608) and its publication in 1638. So obviously it does not appear in the *Dialogue* of 1632. Yet Galileo does make a cryptic remark after the equal arcs proof: he queries whether heavy bodies fall exactly this way and hints that maybe they do not. He writes, "I shall only say that if the line described by a falling body is not exactly this, it is very near to it." That's it; there is no further commentary, but I suspect he is thinking of the parabolic projectile. Why he does not mention it here is rather odd. Perhaps he does not wish to introduce a contradiction here; he is keeping things simple (and circular) in the debate over heliocentricity, which is essentially what the *Dialogue* is all about. Of course, Galileo was confronted with the problem when he eventually published his parabolic law in *Two New Sciences* (1638). Thus we now turn to Galileo's last book, looking particularly at the context of the law of parabolic projectile motion.

Here he first correctly states the law this way: "When a projectile is carried in motion compounded from equable horizontal and from naturally accelerated downward [motions], it describes a semi-parabolic line [i.e., a partial arc of a parabola] in its movement." But a few pages later a question is raised about the termination of this motion. A vertically falling body moves toward the center of Earth, but since a parabolic line "goes ever widening from its axis," the projectile would not end at Earth's center. Galileo maintains that the projectile *must* terminate at the center of Earth, but surely it cannot if its path is parabolic, which (unlike circles and ellipses) is an open curve. Hence, to reach the center of the Earth "the path of the projectile would be transformed into some other [curved] line, quite different from the parabola."

Galileo's answer to this problem is a bit obscure but as I interpret his remarks the point is this: the projectile law is only a *local* law; it does not apply to Earth as a whole. For local phenomena "the distances we employ are so small in comparison with the great distance to the center of our terrestrial globe that [for example] we could treat one minute of a degree at the equator as if it were a straight line, and two verticals hanging from its extremities [also at the equator] as if they were parallel." So locally inertia appears linear, falling bodies move vertically, and hence a projectile appears to trace the line of a parabola. But these are only (local) approximations: for the whole Earth there are neither straight nor parallel lines; and therefore the true law is, not surprisingly, the circular law of motion (that is, rotational inertia). As visually displayed in Figure 8.5, an arc that locally may appear to be a parabola is really globally part of a circle. Just as gravity is an illusion in the larger context, so the parabola morphs into a true circle. As before, everything is subsumed under the rule of the simple and perfect circle, about which, it seems, Galileo exhibited a case of mental inertia. How Aristotelian!

In one sense Galileo was right. The parabolic law is a local approximation—not, however, to the rule of circles but to the law of ellipses discovered by Kepler in 1609, a discovery that Newton fruitfully exploited (see section 10.2) but Galileo chose to ignore—for obvious reasons, though at his peril.

An aside: Galileo's obsession with circles, and the corresponding rejection of ellipses, may have relevance to his attitude toward the art of his time. His allegiance to the more classical artistic style and his denunciation of Mannerism (see section 7.2) is a reflection of this, for the circle was a ubiquitous "classical" Renaissance motif, whereas Mannerist distortions often involved various ovals (shades of Kepler?) and other stretched (anamorphic) shapes. Both in art and science Galileo was fixated on the circular form.

Looking at the history of science anachronistically, earlier discoveries often appear obvious and mistakes seem foolish. But looking at these matters in their context—by attempting to immerse ourselves into the past and even trying to think within the earlier framework—allows us to appreciate the difficulties even geniuses had in extricating themselves from preconceived viewpoints. This intellectual exercise goes a long way to explaining why the breakthroughs of individuals, geniuses though they may be, are usually only partial breaks with the past, and why it often takes the combined ingenuity of several scientists to make a clean break. Galileo, despite his insights, newfangled ideas, and discoveries, was, as seen, very much still ensnared in the circles of Aristotle.

I add two caveats to this chapter. In a letter of 1637 Galileo dismissed his proof of the illusion of gravity, calling it a "jest" and a "poetic fiction" and pointing to his comment in the *Dialogue* that the proof is only a series of "curiosities" (*bizzaria*). But this letter was penned after his recantation before the Inquisition and should be read in that context. The unreasonableness and downright cruelty of the repressive system of house arrest to which he was subject may be gleaned by an example: when Galileo was suffering with a painful hernia and asked to visit his doctors, his request not only was rejected but the Inquisitor told him that any further petitions would result in imprisonment. Accordingly Galileo must have been extraordinarily cautious, especially when writing on scientific matters.

Most scholars are sure Galileo took the proof seriously—I think very seriously. Also, he wrote the 1637 letter while he was working on *Two New Sciences,* which contains the law of parabolic projectiles. Now I have interpreted Galileo's oblique references to differences between local and nonlocal phenomena (with respect to Earth) as implying the circle to be exact and the (local) parabola an approximation. Some scholars, however, have put forward the opposite argument. Recall Galileo's cryptic comment after the proof in the *Dialogue,* namely, "I shall only say that if the line described by a falling body is not exactly this [circle], it is very near to it." The implication here is that (at this time) Galileo believed the parabola is the exact path and hence the circle is an approximation.

Did he later change his mind? And, if so, why did he not clarify it in *Two New Sciences?* An answer to the latter is implied in the letter of 1637: he was compelled to distance himself from direct arguments supporting Copernicus. At most, he would make things obscure, perhaps trusting that the reader would read between the lines. As to the former question, this is how I read between the lines in Galileo's passages in *Two New Sciences*—the parabola is a local approximation to the true cosmological circle.

Real-World Physics: Lost and Regained

Galileo was well aware that the laws he was abstracting from nature would not submit exactly to experimental tests. In a sense Aristotle's physics was more real, since he was concerned with physical objects moving in media, as they do on and about our Earth.

Using an obvious intuitive argument, Aristotle concluded that heavy objects should fall faster than lighter ones, since (holding them in our hands) we can feel the heavier one striving harder to reach Earth. Moreover, the resisting medium retards that motion, so the speed of the fall should be inversely proportional to this resistance. Indeed, his law of falling bodies was qualitatively correct; falling in the same medium, heavy objects do fall faster than lighter ones. If two otherwise similar spheres, say weighing 12 pounds and one-quarter pound, are dropped from a 100-foot tower, when the heavier sphere hits the ground the lighter one will still be about 15 feet away. The problem for Aristotelian physics is that quantitatively this does not fit Aristotle's law, since he (again, intuitively) assumed that their speed of fall was proportional to their weight. If this were so, then when the 12-pound sphere hit the ground the one-quarter-pound sphere should only have fallen a few feet. Does this mean that the theory should be abandoned?

Galileo thought so, and often pointed to such discrepancies as evidence against Aristotle's theory. His solution was to abstract to an imaginary world where things move in a vacuum, so the resistance is zero. Only in this ideal world, he believed, could we find the perfect, mathematical (namely, geometrical) laws of nature. And he did: he deduced that heavy objects fall independently of their weight, that they accelerate as they fall, and that the distance they fall is proportional to the elapsed time squared. But Galileo was under no deception that objects in the real world would obey these laws exactly. That's why some historians contend that Galileo never actually tested his hypothesis by dropping two different weights from the Leaning Tower of Pisa. For decades historians declared it was a myth; I was taught this in graduate school. Stillman Drake, however, has convinced me that Galileo did perform this celebrated experiment to falsify Aristotle quantitatively and to verify his own law as an approximation. This was not so much a controlled experiment as a public demonstration directed to philosophers. The two weights hit the ground close enough to being simultaneous that Galileo could openly conclude that they would do so in the limit of no resisting medium.

Newton and others immortalize Galileo's achievement in the history of science, mainly through their further elaborations and expansions. But what about the physical world of objects moving a medium? Newton groped with this problem in the second part of his *Principia,* but with little success. He only laid the groundwork for what proved to be a formidable and complex matter that was later developed by several brilliant mathematical physicists. Thus, the seemingly simple case of a sphere falling in a medium was not completely solved until the mid-19th century by the Irish physicist George Gabriel Stokes.

There is a delightful irony in his solution. If we write Aristotle's law for a weight falling in a medium as a proportion (although he only worked in ratios), it appears as

$$S \propto W/R,$$

where S is the speed of fall, W is the weight, and R is the resistance of the medium.

Stokes's law applies to a sphere of radius r and density d, falling in a medium of density δ and viscosity v. After initially accelerating briefly, the falling sphere reaches a terminal speed S, due the resistance of the medium, and falls at this constant rate, such that

$$S \propto r^2 (d - \delta) / v.$$

Of course this law was deduced over two millennia after, and is a far cry more accurate than, Aristotle's. Moreover, Stokes's law is based on Galileo's abstract law of falling bodies, although now augmented to taken into account real-world friction of a medium. Nevertheless, note the formal similarity between the two real-world laws. Stokes's law entails a "weight" in the numerator (the density of the sphere) and the resistance of the medium (now, the more sophisticated concept of viscosity) in the denominator. Ironically, it's more like Aristotle's law than Galileo's law.

Notes and References

The law of projectile motion is presented in Galileo Galilei, *Two New Sciences,* trans. Stillman Drake (Toronto: Wall and Thompson, 1989), pp.147–148, 217–218, and 221–225. Stillman Drake and James MacLachlan, "Galileo's Discovery of the Parabolic Trajectory," *Scientific American,* 232 (March, 1975), pp. 102–110.

Galileo Galilei, *Dialogue on the Two Chief World Systems,* trans. Stillman Drake (Berkeley: University of California Press, 1967), pp.164–167 (for a stone falling from a tower), 190–196 (for rotational motion at the equator), and 398–399 (for the ball in a bowl of water experiment).

Galileo's quotation on inertia is found in his "Letters on Sunspots, " reprinted in *Discoveries and Opinions of Galileo,* trans. Stillman Drake (New York: Doubleday, 1957), p. 113. The letter of 1637 is quoted in Stillman Drake, *Galileo at Work: His Scientific Biography* (Chicago: University of Chicago Press, 1978), pp. 376–379.

An alternative interpretation of circles and parabolas is by Ron Naylor, "Galileo, Copernicanism and the Origins of the New Science of Motion," *British Journal for the History of Science,* 36, No. 2 (June, 2003), pp. 151–181.

On Galileo's concept of inertia compare, Dudley Shapere, *Galileo: A Philosophical Study* (Chicago: University of Chicago Press, 1974), Chapter 4 ("Galileo and the Principle of Inertia") with Stillman Drake, *Galileo Studies* (Ann Arbor: University of Michigan Press, 1970), pp. 240–278, and *Galileo at Work: His Scientific Biography* (Chicago: University of Chicago Press, 1978), pp. 376–377.

The quotation about the Arsenal of Venice is by Ingrid Rowland, from her essay, "The Nervous Republic," in *The New York Review of Books* (November 1, 2001), p. 12.

The binocular rivalry experiment is mentioned in John R. Searle, "Consciousness: What We Still Don't Know," *The New York Review of Books* (January 13, 2005), pp. 36–39.

9
Aesthetics and Holism: Newton on Light, Color, and Music

Newton is famous for showing that Galileo's (correct) physics of motion combined with Kepler's three laws of planetary motion are consequences of the inverse-square law of gravity. Like scientists today, Newton siphoned off these three laws from Kepler's life work, leaving behind seemingly remnants of mysticism. Nonetheless, and surprisingly, Newton's mind set was closer to Kepler's than Galileo's. I will illustrate this by looking at Newton's study of light.

9.1. Newton Creates the Spectrum

Science students often make up mnemonics for Newton's seven colors of the spectrum. Memorized and repeated like a chant, the colors are red, orange, yellow, green, blue, indigo, and violet (or the mnemonic ROY G BIV). But, why seven? After all, the spectrum literally displays a *continuum* of colors, one blending into another from the reddish side to the bluish side. So where did the canonical seven come from? In his *Opticks* (1704), Newton admits that the spectrum is a continuum but still delineates the sequence of seven colors. The spectrum, he writes, is "tinged with this series of colours, violet, indigo, blue, green, yellow, orange, red, together with all their interme-diate degrees in a continual succession perpetually varying."

What is not commonly known is that when he *first* studied the spectrum Newton saw only *five* colors. Shortly thereafter he added two more. The story of why he did this, and more, reveals how Newton's thinking about the world, despite its seeming modernity, was move closely aligned with Kepler's.

His systematic study of light began, he tells us, in the mid-1660s when, as a student at Trinity College, he bought some prisms at a country fair. Prisms were toys, artifacts for amusement, like kaleidoscopes; the viewer held them in front of the sun and looked at the rainbow of colors. But Newton turned this toy into a scientific instrument by doing something that rarely was done before: he darkened his room and allowed a beam of sunlight to shine through the prism, thus projecting the colors onto the opposite wall. The few who tried this, projected an image only a few inches. But Newton reports that the wall was 22 feet from the prism with the resulting spectrum of light being 13¼ inches long and 2⅝ inches wide.

Newton went on to perform a variety of experiments with colored spectra. One of the most famous in the entire history of optics he called his *experimentum crucis* (crucial experiment). Using a second prism, he individually cast each specific color onto the wall after the first prism had separated them. The colors did not change: picking up (say) the color green, it remained green even when refracted through the second prism. This experiment is shown in an often-reproduced drawing by Newton (Fig. 9.1), which sometimes mistakenly is thought to be an original sketch from the 1660s. Actually it was made many years later; he sketched it 1721 (in his late 70s) for the second French edition of the *Optiques*. Nevertheless, we assume it records—in general, if not faithfully—what Newton did in the 1660s. This experiment he believed unambiguously disclosed the particle (or discrete) nature of light, since whatever is carrying the color through the first prism is not modified by the second one. Newton writes on the sketch, *nec variant lux fracta colorem* (that is, light does not change color when refracted [the second time]); the phrase appears thrice, although curiously it is crossed out once. (I have not reproduced this text in my diagram.)

FIGURE 9.1. Newton: optical experiment. A sketch of Newton's diagram of his famous experiment on the refraction of light through a prism, from which he deduced the colored division of the spectrum and the inference that light has a particulate nature. Newton's diagram, it should be pointed out, was a reconstruction from memory made many years later. Of interest is his depiction of five distinct colors in the diagram, harking back it his original division of the spectrum.

If light has a wave- or pulse-like nature, which was what his later arch-rival Robert Hooke assumed, then one would expect that as the light is refracted while passing through the first prism, whatever causes the colors to arise would likewise refract the colored light through the second prism and produce another set of colors. Instead something discrete seems to preserve the color as it passes through the second prism. From this point of view the first prism acts like a filter, separating the discrete colors apparently already within the white light from the sun. It is not surprising that posterity has agreed with Newton in calling his experiment "crucial," for it still convinces just about everyone when first confronted with it. At the very least, science textbooks present the argument as ironclad.

In fact, however, everyone did not initially see it that way. When Newton first made known his hypothesis publicly in 1672, Hooke countered with this line of reasoning: assuming the colors to be already in the white light is like saying that sounds are in strings before they are plucked. Rather, white light is a homogeneous entity and the colors are due to modifications to the light made by the prism as the light is refracted; in short, the prism makes the colors. This was the dominant view of light, often a symbol of God, hence its pure and flawless nature. Newton, contrariwise and hence very radically, viewed the white light as heterogeneous (instead, the colors were homogeneous), so that the prism separated the colors according to their degrees of refraction.

From Hooke's point of view Newton's experiment did not confirm anything. Simply because the white beam is modified in passing through the first prism does not necessarily imply that the colored beam should likewise be modified. The prism may act differently on a colored beam than a white beam. Moreover, the property acquired by the now-modified beam after being refracted by the first prism may not be similarly affected by the further refraction through the second prism. The debate over the nature of light was the first of many clashes between Newton and Hooke.

Newton, Hooke, and the Giants

One of the famous quotations by Newton, often interpreted as revealing his humility, is this: "If I have seen further [than others], it is by standing on the shoulders of giants." The metaphor was not unique to him. It has been attributed to earlier writers, perhaps as far back as the Roman poet Lucan. At least it is found in Bernard of Chartres, an early-12th-century theologian-scholar whose admiration of the ancients provoked this comment: "We, like dwarfs on the shoulders of giants, can see more and farther, not because we are keener and taller, but because of the greatness by which we are carried and exalted." Most intriguing in this context are several images on the celebrated stained-glass windows of Chartres Cathedral. The Gothic cathedral was rebuilt (1194–1220) after a fire destroyed the original Romanesque structure. The images depict the four New Testament evangelists literally riding on the

FIGURE 9.2. Window: Chartres Cathedral.
A sketch of St. Luke on the shoulders of
Jeremiah from a stained glass window.

backs of four Old Testament prophets, for example, Luke on the shoulders of
Jeremiah, which may be deemed as visualizations of the famed aphorism
(Fig. 9.2). In Newton's time the phrase is found in the widely read *Anatomy of
Melancholy* (1621) by Robert Burton.

Newton's version appears in a letter to Hooke in response to one from him.
The background to this exchange is a previous acrimonious exchange initiated
by Hooke's attacks in which he accused Newton of plagiarism and errors.
Hooke writes on January 20, 1676, that he wants to correspond privately on

some contentious matters of science, specifically research that he had begun but did not complete, and about which Newton had "gone further" in pursuing. The letter appears to be an attempt at reconciliation. In his reply Newton (February 5, 1676) agrees about the private exchanges and, apparently in response to Hooke's remark about Newton having "gone further" than he, Newton refers to the work of Descartes as providing a basis for optics and also mentions Hooke's work on "ye colours of thin plates" (see section 9.3). Newton then says: "If I have seen further it is by standing on ye sholder of Giants." In light of previous caustic exchanges between them, it has been suggested that the statement is really a put-down—unquestionably a particularly nasty one—diminishing Hooke's work with a concomitant assault on his physical deformity, for Hooke was a very short and hunchbacked man.

In fairness I should point out that Newton's biographer, Richard Westfall, would not agree. He writes, "I do not accept the interpretation that the ["shoulder of giants"] . . . phrase was a deliberate, oblique reference to Hooke's twisted physique. As Newton said once before in regard to Hooke, he avoided oblique thrusts. When he [Newton] attacked, he lowered his head and charged."

9.2. Newton Counts the Colors

When and why did Newton identify seven colors? What do we know precisely? We have no direct documentation of his first optical experiments; the earliest material is dated a few years after the experiments were performed. In the fall of 1669 Newton was appointed Lucasian Professor of Mathematics at Trinity College and he immediately planned a series of lectures on light and color, the subject on which he had been working on and off for about 5 years. These lectures, written between 1670 and 1672, are the earliest sources of Newton's thoughts about light.

In the first lecture he presents his celebrated experiment of projecting a beam of light through a prism in a darkened room. The resulting spectrum he describes this way: "Hence, insofar as the rays are so disposed that some are refracted [by the prism] more and more than others, they generate in order these colours, red, yellow, green, blue, and purple, together with all the intermediate ones that can be seen in the rainbow." (Recall that in the above quotation from the *Opticks* he used violet; throughout his writings he uses "purple" and "violet" interchangeably.) Here he recognizes a continuum of colors, as mentioned later in the *Opticks,* but he also identifies as specific colors only *five,* not seven. So where are orange and indigo? They do not appear until the eleventh lecture, revealing that Newton originally delineated only five distinct colors in the spectrum—or said prosaically, those five are what he initially *saw.*

In the eleventh lecture he goes back to the spectrum, which he now casts onto a piece of paper. He tells us that he marked with a pen the boundaries between the colors; in addition, he marked where each color was the strongest ("the most perfect colours of their kind"). This was followed with a quantification of the spacing of the colors, by dividing the spectrum into 60 equal parts. The resulting ratios are as follows: violet or purple (16/60), blue (14/60), green (10/60), yellow (11/60), and red (9/60). As such, the colors are not equally spaced across the spectrum. He then identifies the colors he sees at the boundaries between the colors as follows: between purple and blue is indigo, between blue and green is sea green, and between yellow and red is orange (no color is designated between green and yellow). This is the first appearance of indigo and orange, and they are introduced as *boundary* colors only. I have reconstructed these observations in Figure 9.3. Newton points out that he made a concerted effort to make sure that he was being objective in these observations; he tells us that he asked "others" to verify that his markings were correct ("trusting not only my own senses"). Moreover, he drew yet another scale figure of the spectrum and "projected the colored light onto this figure to confirm once more whether every color would be confined within the limits assigned," and he found that they all fit as he had drawn it. No doubt Newton was being meticulous in his quantification of the light.

As noted, he was not only interested in the boundaries between the colors, but also marked where each color peaked in intensity. A close look revealed that these "peaks" were not at the centers of their intervals. The brightest part of each color he located as follows (see the horizontal arrows in Fig. 9.3): the "most brilliant" blue was shifted toward the middle of the spectrum; the "fullest" yellow was also shifted toward the center (or green); red and purple, however, were shifted away from the center; and, finally, "green alone sprang up in the middle" of its boundaries. Newton then examined the proportional arrangement of these colors, noting that colors were crowded near the center and spread out near the

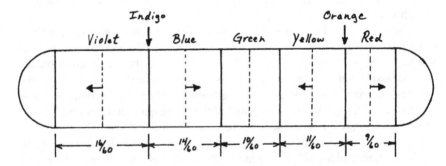

FIGURE 9.3. Newton: colors of the spectrum. A reconstruction of Newton's division of the spectrum, showing his original division of five colors, his quantification of the colors, and the subsequent introduction of indigo and orange as boundary colors. The arrow indicates the shifting of the brightest parts of individual colors, thus opening gaps for indigo and orange to assume full status, resulting in the famous seven-colored spectrum taught today.

edges. As the horizontal arrows I've drawn reveal, there are gaps where indigo and orange constitute boundary colors. At this point Newton fills in these gaps with the boundary colors. This, consequently, is where he made the transformation from the original five to the seven colors of the spectrum. Indigo and orange are transformed from boundary colors to *individual* colors of the spectrum. Why? He tells us: so as "to divide the [spectral] image into parts more elegantly proportioned to one another." After all, indigo and orange are the next brightest ("eminent") colors and by filling these gaps a spectrum of more equal division among the colors is created, so that "everything turns out proportionate to the quantity of green with a more refined symmetry." Listen to Newton's language: *elegance, proportion,* and *symmetry*—aesthetic terms usually associated more with art than science. Nevertheless, we have heard such language from Copernicus and Kepler, among others.

So there it is. It began as a careful observation of five distinct colors. After quantifying the spectrum, Newton filled in what he saw as two gaps with boundary colors, with the resulting now-canonical seven-color spectrum. Sea green, incidentally, remained a boundary color. And if what Newton says in the *Optical Lectures* is what actually happened a few years before—and I cannot find any reason to doubt him on this—then he justified the change for aesthetic reasons.

Henceforth Newton affirmed the seven-color spectrum. Now look again at the famous sketch of the so-called crucial experiment (Fig. 9.1) made in 1721, about 55 years after the first experiments. It delineates only five colors emerging from the first prism. Why only five, when the *Opticks* specifies seven? Was Newton just being careless in his drawing—he certainly was no artist, as was Galileo—or was this an unconscious slip of the pen, harking back to the original experiment performed as a student? There does not seem to be any way of answering this question.

Newton and Indigo

Newton used purple and violet interchangeably but they really are different. Purple has ancient origins, associated with royalty; kings and queens were often clothed in purple. The dye came from the *Purpura mollusk*; hence the word purple, which was sometimes used for a deep red. Violet is a Middle English word vaguely used for the color of the flowers. Indigo was a much more recent word, introduced in English in the mid-16th century. The dye came from a plant (genus, *Indigofera*), obviously imported from India. Apparently it was a popular color at the time. It is a bit odd that Newton would introduce this color into his spectrum, since in his mind God implanted these colors into the sunlight. Why pick a trendy color?

9.3. From Colors to Music

Returning to and continuing with the *Optical Lectures*: having delineated the seven-color spectrum, Newton immediately moves on to something else in the next paragraph. The previous paragraph ended with the seven-color spectrum having "a more refined symmetry." The next paragraph begins, "Consequently . . ."—as if what he is going to say follows logically from the aesthetically motivated spectrum of seven colors. So, says Newton, "Consequently, after these colours had been interspersed, I once more made observations" of the division of the spectral image and found that "the colors were proportional to a string divided so it would cause the individual degrees of the octave to sound." This is what we may call the color–sound analogy, and it needs to be explained. (By the way: this has no relevance to Hooke's plucked string analogy in his critique of Newton's interpretation of the spectrum.)

Music was fundamental to the Western educational system. It was one of the seven liberal arts, which formed the basis of university education from the late ancient world into the 19th century. Specifically "music" was music theory, not performance; indeed it was called "harmony." *Elegance, proportionality,* and *symmetry* were the evocative words used by Newton to justify the expansion of the colors in the spectrum. With the color–sound metaphor we hear Newton extend the language to include the "harmonies of color." An aesthetic phrase, to be sure, but "harmony" can also be taken literally in this case. In fact, it was a subject (music = harmony) he studied as a student at Trinity College. His student notebooks from 1664–1666 reveal his familiarity with various schemes of temperament (see section 6.4). Not surprisingly, the first seven-tone scale that he identifies in the spectrum of light came from his notebooks. The details do not concern us here but it is interesting to look at its form: the intervals are 8/9, 15/16, 9/10, 8/9, 9/10, 15/16, 8/9. Note the symmetry, which makes it a palindrome. In other words, he postulates that the intervals of the seven tones of the musical scale correspond to the intervals of the seven colors of the spectrum.

All this is a bit of a fudge, and Newton concedes as much. The palindrome scale does not quite fit, since the spectrum is not (as we know) perfectly symmetrical, despite Newton's imposition of some symmetry with the additions of indigo and orange. He reports that, in fact, another scale "fits"—a scale that likewise is taken from his student notebooks. Both scales only approximate the data, however; neither fits better. There is thus no empirical reason to choose one over the other. Yet Newton makes a choice; he prefers the palindrome because it seems to reveal a closer affinity between the "concordances of sounds" and the "harmonies of colours." Exactly what he means by this is not clear, but there is a hint in the following parenthetical phrase after "harmonies of colours": he writes, "such as painters are not altogether unacquainted with, but which I myself have not yet sufficiently studied." It seems that something about the way artists use colors contains a clue to understanding this problem.

Now, hold that thought: I want to back up a bit. After enlarging the spectrum, Newton immediately links it to a musical scale, and then draws upon his student notebook on harmony. Why? Why make the analogy at all? What drove him toward such an idea? The answer is found, in part, in other experimental work on the optics of thin films.

Newton learned about the colors of thin films from Hooke, who had revealed a periodicity within the colors. Newton was the first to measure this periodicity, and his later publication of this led to another dispute with Hooke. Newton specifically studied the colored rings in the thin film of air between a (convex) lens and a sheet of glass (or a planar convex lens); he spoke of these—in a phrase that rings poetic to 21st-century ears—as "ye coloured circles twixt two contiguous glasses." Today they are called interference fringes—or, we just call them "Newton's Rings"—and are explained on the wave model of light. Although Newton is most famous for his introduction of the particle model of light, he also used a quasi-wave model to account for certain optical phenomena that did not fit the discrete or particle model. Such was the case for thin films; he attributed the periodicity of the colors to vibrations of the medium. Nevertheless, his was a primitive effort by today's standards toward what the 19th century would discover about the complex nature of the interference of light in different media.

The relevance of this to our story is that Newton believed he discovered in the periodicity of thin films the ratio 2/1 for the "vibrations" of red to violet. Empirically, this numerical ratio was the ratio of the thickness of the air twixt the glasses at the red and the violet rings. Of course, 2/1 is the musical octave! Here, it seemed, was a numerical ratio, fundamental to music (see section 6.4), springing up in an optical phenomenon. This must have set him thinking something like this: the octave is composed of a seven part scale (C, D, E, F, G, A, B, and then C' to complete the "octave"); the optical spectrum contains seven colors; moreover, the same 2/1 ratio as the musical octave is found between the colors red to violet. (Of course, this too is a bit of a fudge, since the 2/1 corresponds to the C'/C cycle, for which there is no corresponding analog in light.) Nevertheless, and apparently, this was the catalyst for the color–sound analogy, which in turn stimulated him to seek in his student notebooks a scale that might fit the spectral data. The work on thin films provided the initial justification for the color–sound analogy, and perhaps even for fudging the ratios to fit. It was certainly an act of intellectual syncretism, if Newton had not already thought of the analogy.

Hertz and Holism

Newton's introduction of orange and indigo into the spectrum of colors was guided by a sense of aesthetics. Sometimes such guidance is fruitful, sometimes not.

The important experimentalist and theorist Heinrich Hertz in 1890 attempted a reformulation of Maxwell's equations for electromagnetism, which contain an asymmetry between the electric charge in one equation (as a source of the electric field) and merely magnetic fields in another equation. (Recall that a magnet is always polarized, with both positive and negative poles; cut a magnet in half and it still is polarized.) So Hertz put forward the concept of a magnetic "monopole" (or magnetic charge)—namely, a positive or negative pole alone. But, however much Hertz's modified equations are symmetrical, experimental science has yet to detect any such sources of magnetism. Magnetic monopoles have yet to be found in nature.

9.4. Newton's Holism

Was Newton's color–sound analogy just that? Was it an accident that colors and sound seemed to fit? In an unpublished letter of 1672 he sketched a hypothesis based on the argument that as air vibrations produce sound, and harmonious sounds are due to mathematical ratios (such as 2/1 for the octave), so light vibrations strike the retina and excite the bodily ether therein, traveling to the brain, and affecting the soul in the same manner as do sound vibrations from the ear. To get around the problem of comparing the seven-part spectrum to the eight-part octave, noted above, Newton includes the rounded ends of the color spectrum, in which there are deep red and deep violet, so that the spectrum may be divided into about eight parts. "To which end I would suppose the vibrations causing the deepest scarlet to be to those causing the deepest violet as two to one; . . . and the reason why the extreames of colours purple and scarlet resemble one another would be the same that causes octaves (the extreames of sounds) to have in some measure the nature of unisons." It is curious to speculate that if Newton had transformed the sea green from a boundary color to an individual color, as he did with indigo and orange, he would have had the necessary eight (octave) colors for the spectrum.

Newton then published this in a paper, "An Hypothesis Explaining ye Properties of Light" (dated December 7, 1675). In it he reiterates that as sound is produced by "vibrations in the air," so light in some cases is "vibrations in the ether." When light strikes the eye the vibrations travel "through the optic nerves into the sensorium," just as sound travels in a trumpet. Both are caused by the same mechanism, namely matter in motion. Thus the connection between color and sound is more than a mere analogy. They are part of the same mechanism. They reveal a holistic/unity of nature.

Einstein spent about half of his life searching for a unity between gravity and electricity. He never found it (see section 1.1). Many of his colleagues thought he was an old fool for pursuing this quest. A young J. Robert Oppenheimer, then a colleague at the Institute in Princeton, declared him to be "completely cuckoo." But Einstein never wavered. About 20 years after his death scientists picked up the quest, and the search for a unified field theory is now the major goal of some of the best minds in physics.

These 17th-century notions about color and music may seem strange today, but frankly they were based on a similar holistic worldview. The extent to which Newton would go in his quest is seen, for example, by his obsession with the whole number ratios of the octave/spectrum equivalence. It seems that a closer look at the measurements of the red/violet thickness of the air between thin films revealed a breakdown in the 2/1 ratio. Instead it was about 14/9. Thus in the *Opticks* he again asserts the unity of the seven colors and seven notes, but now submits that the numbers for the colors relate to sounds as the cube roots of their squares. An example is the extreme case, deep red/deep violet (14/9) corresponding to the octave (2/1). Let's do the math: $[(2/1)^2]^{1/3} = (4/1)^{1/3} \approx 1.587/1 = 14.283/9 \approx 14/9$. Clearly this relationship between the color and sound ratios—as Newton seemingly, and desperately, fiddled with numbers—is not exact, but it is close.

In any case, having "discovered" a connection between color and sound, he carried his quest further. As affirmed in the *Optical Lectures,* he wanted to explore what artists do and how this relates to the color–sound analogy.

This therefore brings us back to Newton's phrase "harmonies of colors." What does it mean? He doesn't tell us in the *Optical Lectures.* But he does in the 1675 paper, where he argues that sound and light are both based on the motion of matter. Since we know that the mixtures of various sounds produce either concord (harmony) or discord, it follows then that there should be a corresponding harmonious or dissonant mixing of colors. Our ears reveal the harmonies of the octave and perfect fifth, for example. What about colors? What are the harmonious and dissonant mixtures? In 1675 Newton presents just one case of each: the harmony of blue (or indigo) and gold (or yellow), and the discord of blue and red.

Where did he get these? Frankly, no one knows; there seems to be no such conceptions of these particular color pairs in any color theory of the time. Later, in the *Opticks* (1704) he merely reiterated the 1675 hypothesis about vibrations of the optic nerve and the harmony of "gold and indigo."

There is, however, an important unpublished document that was a draft for a proposition that never was added to the *Opticks,* perhaps because of its speculative nature, although Newton certainly was not averse to publishing his speculations. This document, I think, is a clue to the source of his color–sound ideas. The argument is based on the analogy between the sequence of colors (red, orange, yellow, green, blue, indigo, violet) and the sequence of notes C, D, E, F, G, A, B, C′, with the seven to eight comparison giving Newton some wiggle room. For sound, two adjacent notes produce a discord. In color, he writes, "For instance green agrees with neither blew nor yellow for it is [only] distant from them but a note or tone above and below[.] Nor doth orange for the same reason agree with yellow or red," since they are all adjacent colors. For sound, C to G is the perfect (concord) fifth. Thus, in color, Newton's continues, "But Orange agrees better with an indigo blew than with any other colour for they are fift[h]s. And therefore painters[,] to set off gold[,] do use to lay it upon such a blew." I assume the latter is an analogy with the concord of the third, say C to E. This explains the published "harmony" of gold and blue (or indigo). In short, the perfect fifth and third in music have correlating analogues in art. But the next sentence contains a puzzle: "So red agrees well with a sky-coloured blew for they are fift[h]s and yellow with violet for they are also fift[h]s." True, both yellow to violet and red to blue are "fifths," yet in the 1675 hypothesis Newton spoke of the discord of blue and red! Perhaps it was a mistake, for recall that in the *Opticks* he mentions *only* the harmony of "gold and indigo."

These unpublished speculations shed some light on how he arrived at his ideas on the harmony and discord of colors, which seem to be unique to him. Why he went no further may be gleaned from the next and last sentence, in which he brings up an important difference between sounds and colors. As I interpret it, sounds are pure (a plucked string produces a single note) but pigmented colors are not pure (green, for example, may be a mixture of blue and yellow). "But this harmony and discord of colours is not so notable as that of sounds because in two concord sounds there is no mixture of discord ones, [whereas] in two concord colours there is a great mixture, each colour being composed of many others."

In the *Optical Lectures,* recall, when Newton introduced the concept of "harmonies of colours," he wrote, "such as painters are not altogether unacquainted with, but which I myself have not yet sufficiently studied." My discussion of "Newton's Holism" is, as far as I know, as far as he went.

The concept of color harmony, of course, is not strange today. Interior designers use variations of it. Moreover, just as the issue of whether musical harmonies, such as the perfect fifth, are innate or learned remains open, so the objectivity/subjectivity of color combinations is open to debate. But one thing we do know—the way colors mix. Most importantly, the way optical (prismatic) colors mix is different from the way pigments of colored powders mix. For pigments the primary colors are red, yellow, and blue; so, for example, mixing blue and yellow produces green; and mixing all colors would ideally (if the pigments were pure) produce gray (although one usually gets brown, due to impure pigments). For prismatic colors the primary colors are red, green, and blue; thus, mixing red and green produces yellow; and mixing all together produces white (like Newton's prism experiment in reverse). Incidentally, Newton's conflation of violet and purple was erroneous; they are distinct, since purple is not a spectral color.

An aside: One of the earliest sources of color theory is Aristotle's idea of two primaries, black and white, with all colors coming from a mixture of them. We do not know much about the actual practice of color mixing from ancient times through the Middle Ages. Possibly little mixing was done among artists, since mainly unmixed colors were used in painting. Color mixing began around 1400, about the time oil painting was invented. Several theories of colors are found in Renaissance treatises on art, which reject Aristotle's concentration on only black and white. There is a four-color theory (red, green, blue, and yellow); a three-color theory (red, yellow, and green); and a five-color one (red, yellow, blue, green, and brown). About the time that Galileo was becoming a celebrity with this telescopic discoveries (1609–13), the modern three-color theory was put forward independently by three scholars. A number of Baroque painters adopted it; we find in sundry paintings by mid-17th century (such as those of Nicolas Poussin) a prominence of red, yellow, and blue. Artisans, especially dyers, were using it as well. Thus Newton, as a student, had access to the correct theory of the mixing of colored pigments.

The profoundly important difference between optical mixing and the mixing of pigments was only fully understood by the mid-19th century. In Newton's time, they were thought to obey the *same* laws. Color was color; light rays and pigments were two sides of a unified vision. This is not a peculiar notion; surely it is not obvious that colored light and colored pigments are intrinsically different. But what *is* astounding is that we find 17th-century experimental confirmation of this erroneous theory. Distinguished scientists such as Newton's English contemporary Robert Boyle performed experiments with colored glassed and "verified" that, for example, optically mixing blue and yellow (using colored glass and prisms) produced green and mixing red and blue produced purple, which is patently impossible! Apparently he saw what he wanted to see. It was not until the mid-19th century, under controlled experiments, that blue and yellow light were mixed producing white.

Another aside: if you wish to see optical mixing, a simple procedure with a television with a picture tube will show you how it works. All you need is a magnifying glass and preferably having the TV hooked up to a VCR or DVD player. Put in the tape or disk and when there is a scene with a variety of colors, press "pause." Now use the magnifying glass to look closely at the screen. You will find little rectangular modules containing various combinations of the primary optical colors. By comparing the modules under the magnifying glass with the resulting colors on the screen, you will see how different optical mixing produces different colors. Thus, for example, when all three colors are on the modules, the resulting "color" is white.

Newton, of course, was convinced of the unity of sound and color, and accordingly affirmed the unity of optical color theory and the way pigments mix. In fact, Newton presented some rudimentary ideas on this as early as the *Optical Lectures*. Recall the eleventh lecture (when he still envisaged the spectrum as composed of just five colors), where he delineates the relative positions of the colors, their boundaries, and the shifting of the color peaks (see again Fig. 9.3, and especially the horizontal arrows). He makes three deductions from this. Blue and yellow are shifted toward the green; this is because mixing blue and yellow produces green. Thus the optical spectrum of prismatic colors reveals something about pigments. Although wrong today, it was, at this point in the argument, a reasonable conclusion within the 17th-century framework. He then observes that the red and green are shifted away from the yellow; this is because red and green "do not compound yellow well." Needless to say they do not; mixing red and green certainly will not make yellow. Lastly he points to the purple and green shifted from the blue; this is because purple and green do not "compound blue well." True too. And so he concludes, "Whence the reason is clear why" pigments combine as they do, which is surely a grand leap from the meager amount of data entailed in these three cases. Indeed, if, after introducing orange as an individual color, Newton had looked again at the evidence from the spectrum, he would have noted this: yellow and red are shifted *away* from orange. But a mixture of yellow and red do produce orange; thus they should—according to the logic of the previous argument—be shifted *toward* each other, not away as they are. If this fact does not falsify the theory, at least it raises an important empirical counterexample. This should have been a warning to Newton about his conflation of pigments and optical colors. But as far as I know, Newton never noticed this—at least, he never commented on it. Is this yet another example of a silence by a scientist (see Chapter 4)? Newton did, however, concern himself with the disparity between only three primary pigments in contrast to seven optical colors, but he never doubted the unequivocal unity among the colors and sound.

Ultimately, his certitude prevailed. Newton, like Einstein, was "completely cuckoo" about a unified/holistic view of nature.

Bach, not Schubert

A recent dramatization from the BBC of Einstein's search for a unified field theory is titled "Einstein's Unfinished Symphony." But this is a poor musical analogue: Franz Schubert's Eighth Symphony was not unfinished because he died while writing it. A better analogue might be J. S. Bach's last fugue.

There are various arguments as to why Schubert's symphony has only two movements, but the key point here is that it came within his career; written at the age of 25, he would complete another symphony six years later in the year of his death. Thus the Eighth Symphony was not terminated by death.

Bach's last fugue was: it literally stops in the middle of a musical phrase—a stark thing to hear. For this reason, it is seldom performed or recorded. As with Schubert's work, there are hypotheses as to Bach's possible intended abrupt ending, which I will leave for musicologists to debate. Suffice to say I wish to make one point, which I admit may be a bit of a stretch. Einstein's unification quest, which was not fulfilled and ended abruptly with his death, was very personal, since he was virtually the only scientist pursuing this problem at the time. Bach's last fugue also bears his personal stamp. The main theme is based on these four notes: B-flat, A, C, B-natural. In the German notation of the time, B-flat was written as just B, and B-natural was written as H!

Notes and References

More details on this topic are found in David Topper, "Newton on the Number of Colours in the Spectrum," *Studies in the History and Philosophy of Science,* 21, No. 2 (1990), pp. 269–279. Also useful is Alan E. Shapiro, "Artists' Colors and Newton's Colors," *Isis* 85 (1994), pp. 600–630, although we disagree on the sequential roles of aesthetics and music in the development of Newton's thoughts on light. My remarks on sea green in the spectrum support my view that the musical analog was not the source of the seven-color division, but a consequence.

The primary sources are Newton, *Opticks* (New York: Dover Publications, 1952 reprint of the fourth English edition, 1730). *The Optical Papers of Isaac Newton: vol. 1, 1670–1672,* A.E. Shapiro, ed. (Cambridge: Cambridge University Press, 1984) and *The Correspondence of Isaac Newton, vol. 1 (1661–1675),* H.W. Turnbull, ed. (Cambridge: Cambridge University of Press, 1959). The important unpublished document on the color–sound analogy is in the *Optical Lectures,* p. 546, note 27.

See Richard S. Westfall, *Never at Rest: A Biography of Isaac Newton* (Cambridge: Cambridge University Press, 1980), p. 274, footnote 106, for his comment on the "shoulder of giants" sentence.

Oppenheimer's remark about Einstein is from a letter to his brother dated January 11, 1935, see *Robert Oppenheimer: Letters and Recollections,* A.K. Smith and C. Weiner, eds. (Cambridge, MA: Harvard University Press, 1980), pp. 189–191.

In fathoming much of Newton's optical speculations, I gratefully acknowledge the helpful correspondence with Yaakov Zik, Tel Aviv University.

10
Missing One's Own Discovery
Newton and the First Idea
of an Artificial Satellite

Newton's picture of how successive projectiles launched from Earth can eventually orbit our planet is arguably the most famous diagram in the history of physics. It comes from a sketch originally drawn by him (Fig. 10.1). A later engraving of the sketch (Fig. 10.2) appears in innumerable textbooks and thus has been seen by countless student over the years. But, surprisingly, Newton himself never saw this engraving!

At first this may seem both strange and puzzling. How could he not see it, if it appeared in his *Principia?* The answer to this is elementary: contrary to many erroneous citations, the diagram does not come from that book. The reason why it is not there and why he did not see it is part of the history of the writing of the *Principia.* Also, in telling this story, there arises the possibility of little lie by Newton—or at least a conscious deception.

The picture begets another story, a tale that ultimately questions the extent of Newton's understanding of his own discovery. It seems that his interpretation of the diagram did not entail the physics of motion that we derive today; amazingly, he may not have realized the full significance of what he drew.

10.1. The *Principia* Project: Origin and Execution

How Newton came to write his masterpiece begins with a discussion among three Englishmen sometime in 1684: Edmund Halley, the astronomer and mathematician (later one of Newton's few friends); Robert Hooke, the ingenious polymath of the time (one of Newton's key adversaries); and Christopher Wren, a mathematician and architect (more famous today for the latter profession). The three were mulling over a model of celestial physics. They all agreed, probably based on Hooke's persuasion, that the planets move about the sun in elliptical orbits through a combination of linear inertia and an attractive force between the sun and the planets. Moreover, by an analogy with the inverse-square law for light (see section 2.1), they also concluded that this attractive force might likewise obey an inverse-square law. What they could not do, however, was actually derive this as a mathematical deduction. Isaac

FIGURE 10.1. Newton's original projectile sketch. A sketch of Newton's drawing found in his manuscript at Cambridge University (MS Dd.4.18 f.1v). The circle is a mere 38 mm in diameter.

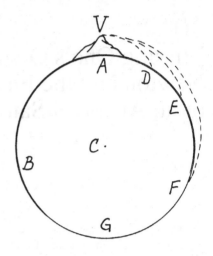

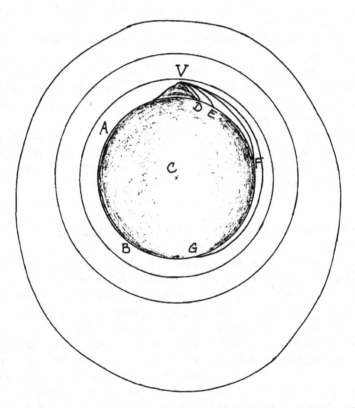

FIGURE 10.2. Published engraving from Newton's sketch. A sketch of the engraving made of Newton's projectile diagram, published posthumously in *On the System of the World*.

Newton's name was mentioned in the ensuing discussion and Halley agreed to go up to Cambridge to consult the young scholar.

Supposedly Halley went in August, met with Newton, and asked him this question: "What would be the orbit of a planet that was attracted to the sun by a force inversely proportional to the square of its distance from the sun." Newton replied, "An ellipse." When the astonished Halley asked how he knew, Newton said he had calculated it. This meant, thought Halley, that this young man had solved a key problem that had eluded some of the best minds of the day. Naturally, Halley asked to see the solution. Amidst his cluttered milieu Newton could not find it. But he promised to send it promptly.

The solution arrived by post in November in the form of a short treatise, written as series of lectures, which Newton called *De Motu* (On Motion). The timing raises a number of questions. Why did it take Newton so long to find the solution or to calculate it again? Why did he not just send the solution? What was the point of writing a longer and more formal treatise?

One thing we do know: this set in motion the writing of the *Principia*. Halley prodded and eventually convinced Newton to expand *De Motu* into a longer treatise on the physics of motion. Newton began the project late in 1684, working seemingly without stopping, especially for two years starting in the spring of 1685. The *Principia* was published July 1, 1687.

What about the time gap between Halley's visit in August and the receipt of *De Motu* in November 1684? One possible answer is found by comparing the first two editions of the *Principia*. The calculation discussed with Halley, namely deducing the elliptical orbit from the inverse square law, does not appear in the first edition. Instead Newton deduces the latter from the former. It turns out that the mathematical deduction of the inverse square law from the elliptical path is a much easier calculation. Newton obviously was aware of this deficiency in his book, for in the next edition (1713) he added a section making the former calculation, which was the original query of Halley.

But there is more: some historians contend that Newton's solution to Halley's query in the second edition of the *Principia* still is not sufficient, that the proof is not correct, and that he therefore did not fully complete the deduction. Defenders of Newton disagree, arguing that the gaps in Newton's proof were intuitively obvious to him. This debate is not yet settled, but one thing does emerge: it seems clear that Newton did not have the proof of the original problem (deducing the ellipse from the inverse-square law) when Halley visited him in 1684, if he ever had it.

In further defense of Newton, the case has been made that Halley did not actually pose the original problem. Undoubtedly Halley made such a visit, but since we know of it from a second-hand account, we cannot be sure of the specific question Halley asked. Even if it were the easier problem (deriving the inverse-square law from the ellipse), which Newton did solve in *De Motu*, there still arise unanswered questions. Why was there a time lag in sending his answer? Why did he send a short treatise rather than just the solution? And indeed why could he not just show the solution to Halley at the time of the visit?

Accordingly, the evidence points more toward Newton being confronted with the original problem. After all, he did attack this problem in the second edition of the *Principia*. A possible scenario may have played out this way: when Halley posed the question, Newton realized the significance of the problem and let his ego get the best of him. He thus boasted that he had solved it, whereas he had not. Put bluntly, he lied. He thus put himself into the situation where he was forced to come up with a solution. We know he did not solve the original problem at this time, despite apparently making a concerted effort. This would explain the 3-month time lag between Halley's visit and *De Motu*. Actually that gap in time may be even longer, for some historians place the meeting with Halley as far back as May. In either case, whether 3 or 6 months, the time gap may explain why Newton sent Halley the short treatise rather than the actual solution. Unable to solve the original problem, Newton could not admit defeat, especially having smugly said he had solved it, and so he cloaked the easy solution in a treatise on motion, hoping that among all the sundry calculations, the reader would not notice that the requested solution was not really there.

It seemingly worked on Halley, who not only was the catalyst for the writing of the *Principia* but also footed the bill. Although the Royal Society agreed to publish Newton's manuscript, it had no money for such work, having just spent its budget on a lavish natural history tome, with numerous expensive illustrations, on fishes. Halley hence paid for the publication out of his pocket—what a friend! (A rather amusing sidelight: due to its financial woes, the society often paid Halley's salary, as clerk, in copies of the fishes book; they tried the same with Hooke, who was secretary, but he refused, accepting only cash.)

We do not know the exact historical sequence of the writing of the *Principia*, but we can reconstruct some of it from the letters Newton sent to Halley keeping him informed of his progress. His first plan was to divide the work into two sections (not separate volumes) called "books." Book I was to be an expansion of *De Motu*, a mathematical study of the motion of bodies under forces in empty space. Book II was to be a popular account applying the physics of Book I to the heliocentric model of the sun and the planets. That Newton conceived of and wrote a draft of Book II is intriguing; he clearly thought that the common reader was an important audience for his ideas. Perhaps he was thinking of Galileo, who wrote the first popular work on science, *Sidereus Nuncius* (1610), announcing his discoveries with the telescope. Much later Einstein would write a short book explaining his theory of relativity to a lay audience.

But something happened to change Newton's mind. The popular account was replaced by a detailed mathematical treatment of the motion of bodies applied to the solar system, although he did incorporate large parts from the popular draft. This section, which he called "The System of the World," became Book III, not II; the reason was that he also added another (different) section, Book II, on the motion of bodies in resisting media between the original Book I and the "System." Some historians think these changes had their origin in a quarrel with Hooke. Whatever the reason, the final *Principia* (published as one volume) was composed of three sections (books) thusly: I, The Motion of Bodies; II, The Motion of Bodies (in Resisting Media); and III, The System of the World (about one third of the latter is about comets; see section 11.2). (Incidentally, the term *solar system* was coined later by Newton's friend John Locke.)

The draft of the popular account Newton deposited in Cambridge University Library. When? We do not know. Possibly as early as the autumn of 1687; at least it was there when he left Cambridge for his civil service job at the Mint in London in 1698. This draft remained in the library, and unpublished, until his death in 1727. A year later the executors of his estate retrieved the manuscript from the library and published it. They bestowed upon it the unfortunate title, *A Treatise on the System of the World,* using the same phrase as the third section of the *Principia,* and thus created the muddle and confusion with that book ever since. They also spawned the confusion over the famous diagram, because it was for the popular account that Newton drew it. Newton's sketch appears only there, and hence it is not in the *Principia.* Surely one reason why the famous diagram is often erroneously cited as being from Newton's *Principia* is the similar title of the published popular account to that of Book III. And, of course, all this is why Newton never saw the engraved version. When he deposited the manuscript in the library it was, it seems, the last time he saw his hand-drawn sketch, and it unquestionably was the only version he ever saw.

Ex Ungue Leonem!

Newton loved mathematical challenges. His niece tells this marvelous story.

The Swiss mathematician Johann Bernoulli once posed a problem directed to all the mathematicians of Europe. Newton learned of it through the Royal Society. The problem was to find the curve followed by a bead falling down a frictionless wire in the shortest time. Today this is a classic problem called the brachistochrone problem (from the Greek for "shortest time"). At the time Newton was working at the Mint, after relinquishing his academic position at Cambridge, and had seemingly given up pursing science. Indeed, it has been speculated that Bernoulli was really directing the problem specifically to Newton, testing the stamina of his mathematical acumen. The day he received the challenge he came home tired from a very exhausting day at the Mint. Yet he spent the night working on the problem, which he solved before morning. The solution, which he sent to the Royal Society, was printed anonymously. But upon seeing the form of the solution Bernoulli knew the author and thus exclaimed: *Ex ungue leonem!* "By the claw, we know the lion."

10.2. Newton's Sketch—and the Problem

I first saw Newton's original sketch of the projectile path (Fig. 10.1) in the wonderful book, *The Ring of Truth,* by the late Philip and Phylis Morrison. Although this was the only place I had ever come across Newton's own drawing, I naively assumed its whereabouts were well known, so I sent a letter to the University Library at Cambridge inquiring about obtaining a copy. The sketch of this most famous diagram I thought would have pride of place in their library. To my surprise, a librarian replied that he had no idea where it was. Accordingly, I wrote to Professor Morrison, who kindly went through his files and found the manuscript number. When I sent this to the library, the staff forwarded it to Mr. Adam Perkins, Royal greenwich observatory (RGO) archivist in the Department of Manuscripts, who subsequently sent me the following description of Newton's manuscript and made arrangements for the copy of the diagram I requested.

The manuscript is approximately 203 mm by 306 mm, and 10 mm thick; it is bound between two boards, with the title, "IS:NEWTON *De Motu Corporum* Lib:II" on the front board, betraying its original aim as Book II of the *Principia.* On the first page of the manuscript Newton has again written, "*De Motu Corporum Liber Secundus,*" but "*Secundus*" is crossed out showing its subsequent rejection as the second *(secundus)* book *(liber)*. On the back *(verso)* of the first page, which is otherwise blank, and near the bottom is the small diagram drawn by Newton (Fig. 10.3).

Looking rather desolate on that empty page, it is his original sketch. The diameter of Earth is only about 38 mm. Why Newton drew such a miniature diagram is a mystery.

The engraved version of the diagram (Fig. 10.2) is reproduced today in countless textbooks, for it succinctly illustrates the range of a projectile and its potential continuum into a satellite. A simple way to see this is to conceive of Earth as being transparent (which is analogous to its mass being concentrated at its center), and then think of the circumference of Earth as a dotted line. The projectile is shot horizontally from a mountaintop, and through a combination of inertia and gravity, it falls to Earth. As Galileo discovered, assuming no resisting medium and for the local approximation of a flat Earth, the path is a parabola (see section 8.2). Looked at globally, however, the path is an ellipse. The relationship between the projectile and Earth (its mass concentrated at the center) is analogous to a planet and the sun. According to Kepler, a planet orbits the sun in an elliptical path, with the sun at one focus. As drawn in Figure 10.2, the projectile's path is also an ellipse, with the center of Earth as a focus. The difference, of course, is that the projectile does not complete its elliptical path; instead it hits Earth at points D, E, F, and G. The elegance of Newton's diagram drawn this way is brought out when we look at the symmetry among the potential closed paths. Consider the projectile hitting point G: if the initial speed is then increased by an infinitesimal amount, the projectile will just glance off the surface of Earth at G and return to the mountaintop, completing an elliptical

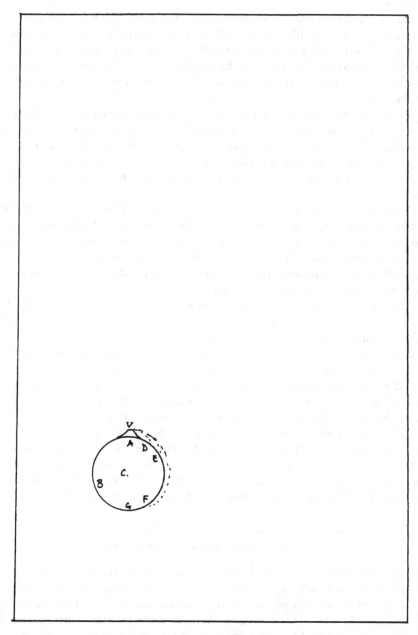

FIGURE 10.3. Page from Newton's manuscript. Copy of the entire page from Newton's manuscript illustrating the scale of the projectile drawing (see Fig. 10.1). The rectangle measures approximately 203 by 306 mm. Why, pray tell, did Newton draw a such a small circle?!

orbit, and remain in orbit as a satellite. This also means that it is impossible to fire a horizontal projectile to hit point *B*, unless, of course, it is fired in the opposite direction. (Alternately, *B* can be reached by firing the projectile at an angle, which the reader may wish to verify by drawing such an ellipse on a transparent-Earth diagram. But our concern here is only with Newton's case of *horizontal* projectiles.)

In conclusion, the maximum range for a horizontal projectile fired from a mountaintop at the North Pole is the South Pole. Any further increases in initial speeds put the projectiles into permanent orbits. Did Newton know this? It certainly is one of the key deductions we make from his diagram today. It also seems to be embodied in his drawing, since the projectile to point *G* is the last one he drew.

A close look, however, reveals that the projectile to *G* is only drawn in the engraving, which was not seen by Newton. In his original sketch he drew projectiles to *D, E,* and *F.* Seemingly he conceived of a projectile to *G*; indeed, he marked the point. Yet he also marked *B,* so did he conceive of a projectile reaching there? Or is *B* denoting something else? The diagram alone is rather cryptic. Perhaps the accompanying text will clarify the matter.

An aside: there may be a simple reason why Newton never drew path *VG*. If you try to reproduce this diagram to scale with a compass, you find that path *VG* is extraordinary close to *VF,* and realizing that Newton was using a 17th century ink pen, the two paths would easily merge or smear together. By enlarging the figure you also discover that he not only used his compass to draw the round Earth but, not unexpectedly, he drew the paths, too, with the compass. Thus *VD* and *VE* are arcs of circles, not ellipses. He began *VF* with an arc of a circle but extended point *F* about 8° further. This left little room for path *VG*.

The engraver also did not draw ellipses, which is readily verified by enlarging the figure and studying its geometry. The outer oval must have been drawn free-hand; the next two are circles, and the projectiles hitting the Earth are also free-hand drawings. Note that the engraver avoided Newton's problem by moving *VF* 90° from *V,* thus leaving enough space to draw path *VG*.

Newton, Cavendish, and Newton's Laws

Just as Newton never saw the famous diagram of projectile motion, he never saw the just as famous equation, $F = ma$. Although this equation is synonymous with Newtonian physics, perhaps surprisingly it is not found in the *Principia.* In fact, it is nowhere in Newton's writings. The brilliant Swiss mathematician Leonhard Euler first presented this modern form of Newton's second law in 1750. In the *Principia,* Newton denotes parameters as ratios or proportions. He thus presents his law of motion as this statement: "A change of motion is proportional to the motive force impressed and takes place along a straight line in which that force is impressed." To see how this relates to the law as we know it, the reader must refer back to a (previous) definition of motion: "Quantity of motion is a measure of motion that arises from the velocity and the quantity of matter jointly." Hence a change of motion entails a change of velocity (namely acceleration), which, along with the quantity of matter ("mass"), is proportional to the force.

Newton also never presented his law of gravity in the format found in textbooks today, namely, $F = GmM/r^2$. The proportionalities between gravitational force and mass he presented in the *Principia* in Book III, Proposition 7, Theorem 7; in Corollary 2 he included the inverse-squared relationship of gravity and distance. At the end of the *Principia,* he summarized these propositions this way (I quote from the General Scholium to the third [1726] edition): "This force [of gravity] arises from some cause that penetrates as far as the centers of the sun and planets without any diminution of its power to act, and that acts . . .in proportion to the quantity of *solid* matter, and whose action is extended everywhere to immense distances, always decreasing as the squares of the distances."

Presented in this manner, the law does not entail the constant of proportionality G, usually referred to as "big G." Instead, the English scientist Henry Cavendish (1731–1810) is usually credited in textbooks as having been the first to measure G. It is true that he used a torsion balance for a series of very precise experiments on the attraction of masses (the celebrated Cavendish experiment); performed at age 67, from 1797–8, they were his last published experiments. But in these celebrated experiments he was not seeking a determination of the constant of proportionality. Like Newton, Cavendish worked in ratios and proportions; moreover, the title of his publication betrays its real objective: "Experiments to Determine the Density of the Earth," or "weighing the world," as he usually called it. In fact, performing the Cavendish experiment to measure "big G" was not done until the late-19th century. By then, physical laws were expressed as algebraic equations and a unit of force (the dyne) was invented.

10.3. The Projectile Path: What Did Newton Know?

As we know, Newton's diagram was meant for the (unpublished) popular *System,* so let us begin with Newton's description of projectiles from that text, our first documentation of his reflections on this. The key section is this:

Let *AFB* represent the surface of the Earth, *C* its center, *VD, VE, VF,* the curve[d] lines which a body would describe, if projected in an horizontal direction from the top of an high mountain, successively with more and more velocity. And, because the celestial motions are scarcely retarded by the little or no resistance of the spaces in which they are performed; to keep up the parity of cases, let us suppose either that there is no air above the Earth, or at least that it is endowed with little or no power of resisting. And for the same reason that the body projected with a less velocity, describes the lesser arc *VD,* and with a greater velocity, the greater arc *VE,* and augmenting the velocity, it goes farther and farther to *F* and *G*; if the velocity was still more and more augmented, it would reach at last quite beyond the circumference of the Earth, and return to the mountain from which it was projected.

Does this text convey a clear comprehension that point *G* is the maximum range of a projectile before it goes into orbit? The answer, unfortunately, is not apparent. We need to study the text in detail. Start with point *G*; even though Newton did not draw the projectile to *G,* he clearly conceived of it since it is mentioned in the text. What then of point *B*? It is not referred to in the text as a target point of the projectile but only as a point delineating the surface of Earth; however, that does not necessarily eliminate it as a target, since point *F* is also first cited as surface point but then is a target, too. There is no clear answer to our question, yet.

Next look at the sequence of projectiles. First is *VD*; then, by augmenting the initial velocity, there follows *VE, VF,* and *VG.* Having now shot the projectile to the South Pole, we reach the last crucial clause in the quotation. Newton writes, "If the velocity was still more and more augmented [why not just one 'more augmented'?], it would reach at last quite beyond the circumference of the earth [that is, it would not hit the surface of the earth], and return to the mountain [top?] from which it was projected [if so, it would go into orbit]." The key phrase is "still more and more augmented." We know that, after hitting point *G,* only *one* infinitesimal augmentation is needed to put the projectile into orbit; thus the extra augmentations seem suspiciously redundant, as if Newton is thinking of further augmentations of the initial velocity resulting in the projectile hitting points such as *B,* between *G* and the mountaintop (note: this could also include returning only to the *bottom* of the mountain). In these cases the projectile would be spiraling toward Earth (as it would if there were a resisting medium) rather than moving in an ellipse. It may or may not be consequential, but note that nowhere does he mention that the shapes of the paths should be ellipses. Of course this may have been obvious to Newton, and the phrasing was solely a matter of style. Sometimes he merely spoke of the planets moving in "circles" about the sun. At most we can say that there is a potential ambiguity between the diagram and the text in the *System.*

Next are the three editions of the *Principia.* Interestingly, the projectile passage is not in the first published edition (1687). Rather, we next find it in Newton's own annotated and interleaved copy of the first edition. (After the publication of the first edition, Newton had his copy interleaved with blank pages facing the original pages, so he could add corrections and additions on these sheets.) His version of the projectile's paths in his annotated copy was then published in the second edition of the *Principia* (1713). It is a variation of the passage from the *System*:

If a lead ball were projected with a given velocity along a horizontal line from the top of some mountain by the force of gunpowder and went in a curved line for a distance of two miles before falling to the earth, then the same ball projected with twice the velocity would go about twice as far and with ten times the velocity about ten times as far, provided that the resistance of the air were removed. And by increasing the velocity, the distance to which it would be projected could be increased at will and the curvature of the line that it would describe could be decreased, in such a way that it would finally fall at a distance of 10 or 30 or 90° or even go around the whole earth *before it falls,* or, lastly, *so that it might never fall to the earth but* go off into the heavens and continue indefinitely in this motion. [The reason for the italicized phrases, by me, will be explained below.]

The military metaphor probably speaks to significant improvements in cannons at the time.

The similarity between this and the passage in *System* is striking. In both, the increasing initial velocities produce longer paths, hitting Earth further and further, until the projectile finally goes into orbit. There is no specification of the paths; in the *Principia* he speaks of the "curvature of the line"—a strange phrase if he is thinking of ellipses, but which would likewise embrace spirals.

What of the differences, do they cast any light on our question? In the *Principia* he writes of the lead ball going from 10° to 90° but does not include 180°. Recalling that no figure accompanies this text, this means that nowhere is Newton mentioning this crucial case. It would seem that *if* he knew this was the last strike on Earth before the ball went into orbit *then* he would isolate this case. Moreover, after specifying the 90° case, and before mentioning that the projectile goes back to the mountaintop and hence into orbit, Newton speaks of the ball going "around the whole earth before it falls." Note that he does not say *halfway* around Earth. Therefore, there seems to be only one way to interpret this: that the ball spirals around Earth and lands near the base of the mountain, since it both falls and circles the entire Earth.

This, of course, is mathematically wrong. There is no such spiraling of a projectile in a vacuum. Spirals only result when a medium is introduced. If Newton is thus speaking of a spiraling motion, it would means that he is not aware of point G (the South Pole) as the maximum projectile target.

Regarding the italicized phrases in the quoted passage, this brings me to the third edition, for these phrases Newton deleted from the last edition (1726). At first blush this seems to show that Newton caught his error; he realized that there was no spiral path, and thus eliminated the phrase about "falling." But a closer read reveals otherwise. There is still the case of the ball going "around the whole earth" followed by the case of its going into orbit. These two cases have not been linked together, and, importantly, there is no indication of any knowledge of the maximum target of 180°. At best it seems that Newton made these changes to eliminate what he saw as merely redundant phrasing. So the third edition of the *Principia,* written shortly before he died, does not contradict my assertion that Newton was unaware that the South Pole was the maximum range of a projectile shot horizontally from a mountaintop at the North Pole.

Although physicists surely know that all such projectiles are ellipses, some historians have made Newton's error. I have found quite a few books (and, now, Internet sites) that reproduce a version of Newton's diagram with projectiles spiraling from the mountaintop to points further than 180° (such as Fig. 10.4). It is an easy mistake to make, by extending Newton's thought experiment, say, this way: Start with Galileo's law of a (local) parabolic projectile; increase the initial velocity so that the projectile is shot further and further; inductively one constructs a series of spirals

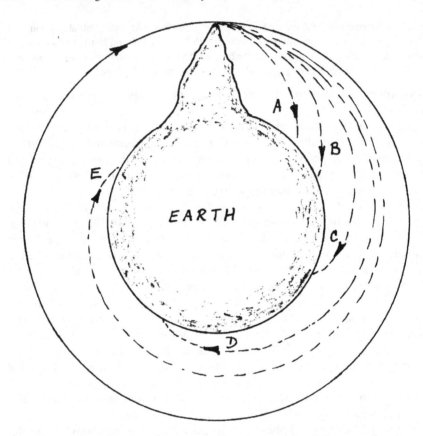

FIGURE 10.4. Erroneous diagram of projectiles. An example of the type of erroneous diagrams of projectiles shot horizontally from Earth as found in various textbooks. I should point out that all the examples I found were in history of science books, not science textbook. The caption to the diagram I have copied reads: "Normally, a projectile (A) will soon fall to the ground. But, argued Newton, if its speed is increased sufficiently (B, C, D, E ...) it will eventually circle the Earth completely."

around Earth until the crucial velocity is reached when it goes into orbit. This is one reason some historians may have drawn the wrong conclusion and hence the inaccurate diagram. Alternately, or even concurrently, they too may have read Newton's text as indeed describing spiraling paths. The more important issue, however, is whether Newton ever was aware of his error, if (as I believe) he actually made it.

A reader may ask: Could not these tedious and detailed analyses of Newton's every word ultimately be irrelevant and meaningless, not really penetrating the mind of the writer? I think not: Newton was meticulously careful with words, persistently editing and re-editing his writings. My answer to the above question is, therefore, no.

The argument presented above—that Newton made this error and died without realizing it—was published in 1999 by me and a colleague in physics, Dwight Vincent. When subsequently several Newton scholars promptly challenged it, we were not surprised. One must be careful when knocking heroes off their pedestals, and Newton, especially, time and again will have his defenders. What began as a series of cordial email exchanges resulted in the publication of a defense of Newton by Michael Nauenberg of the University of California at Santa Cruz, which was followed by our rejoinder.

As I see it, the main challenge by Nauenberg to our thesis is his pointing out (correctly!) that the *Principia* actually contains the mathematical formulation that the South Pole *is* the maximum target point. It appears in Book I, Proposition 45.

Before presenting Newton's exposition, let me remind the reader that Newton's geometrical approach to solving problems in dynamics is a far cry from the analytical method found in textbooks today. Just as Newton probably would comprehend little in a textbook on Newtonian mechanics today, however strange that may sound, only scholars steeped in 17th-century science can read much of the mathematics of Newton's *Principia*.

Here is the formulation: For a projectile under a central force ($1/r^n$) launched perpendicular to the radial direction, where Θ is the polar angle relative to the initial radial direction, the formula from Proposition 45 (as Nauenberg writes it) is

$$\Theta = 180° / \sqrt{3 - n}.$$

(Actually, even this is a simplified and modern version of Newton's much more cumbersome notational system.) Thus for $n = 2$, that is, the inverse square law (hence, gravity), $\Theta = 180°$. So Nauenberg is right: the fact that the South Pole is the maximum range for the horizontal projectile *is* in the *Principia*.

Does this therefore falsify our argument? Maybe, but not necessarily, because there is still one more question to be answered: Was Newton aware of this? This question is not as odd as it may seem at first. That a scientist could be aware of some consequence of his work, or even should be aware, is no guarantee that he is aware. James Clerk Maxwell, for example, died of cancer at the age of 48 without realizing the full significance of his theory, particularly the production and detection of electromagnetic waves. He certainly could have realized this, since it was entailed in his equations. But he did not. A truism from the history of science is a mantra of this book: often things are clear only in retrospect.

Let's look at the context of Newton's formulation of the projectile problem. What does he say about it? In Proposition 45, after deriving the above *general* formula, he considers three specific "examples": where r is constant ($n = 0$), an inverse force ($n = 1$), and $n = 11/4$. Note that *none* of these apply to gravity and hence the projectile problem. The apparent reason for his interest in these other forces—rather than the inverse square force ($n = 2$), which we ascertain *should* be his focus of attention—seems to be that he was especially curious about the geometry (not the physics) of the various "examples": thus, for $n = 11/4$, $\Theta = 360°$.

So where in the *Principia* is the specific (and *explicit*) result that the South Pole is the maximum target point? In the course of deriving the previous examples,

Newton mentions *in passing* that "a body in an immobile ellipse . . . completes the angle *VCP* [i.e., Θ in the above notation used by Nauenberg] (so to speak) of 180°." Stop; read that phrase again, because that's it! Again: "a body in an immobile ellipse . . . completes the angle *VCP* (so to speak) of 180 °." I'm sure only a very attentive reader of the *Principia* would notice that *this* is the case we are pursuing. One reason it is not transparent is that the reader must remember, from a *previous* derivation in Proposition 44, Corollary 2, that the "immobile ellipse" entails the inverse-square law! Without recalling this, the significance of the quoted phrase is lost. And Newton does not help, since he does not point this out. Hence there is no clear indication that this *short* phrase buried within a *longer* sentence about a *different* example, is the case of a projectile fired halfway around Earth. Moreover, why he uses the parenthetical phase "so to speak" remains a mystery: the angle *is* 180°. Is Newton confused about this? Is this further reason to doubt his awareness the consequences of his formula? One thing I can say: I am not convinced of Nauenberg's argument.

Most importantly, and furthermore, there is nothing else to show unequivocally that Newton fully understood the physical significance of his derivation. One would think that if he did, then at least by the third edition of the *Principia*, when editing the projectile passage, he would have changed it to something like this: " . . . until at last it should fall at the distance of 10, 30, 90, or 180°, or even might go quite round the whole earth." But in the absence of a definitive statement of this, coupled with the implied spirals in the projectile passage, I maintain my skepticism about Nauenberg's challenge.

Accordingly, I assert that although Newton could and should have known that the South Pole is the maximum target for a projectile fired horizontally from the North Pole, he did not. At least, nowhere does he present explicit evidence of a physical awareness of this most profound deduction from his theory.

Until further evidence is forthcoming, the dispute seems to be at an impasse. I will close, however, with one more argument. In the *Principia* Newton demonstrates how the inverse-square law of gravity accounts for a two-body system, such as the sun and a planet. He also introduces what today we call the three-body problem. As he conceived it, the regular motions of any three bodies (involving, say, the sun and two planets or the sun–Earth–moon system) would be almost impossible to preserve due to the unstable equilibrium among their mutual gravitational interactions. Hence he believed that God intervenes, keeping the solar system from degenerating into chaos.

Now if Newton indeed knew that a horizontal projectile that reaches the South Pole requires only a small increase in its initial speed in order to go into orbit, he could have made the deduction that these orbits are much more stable than he otherwise assumed. Such knowledge would have saved him considerable fretting over unstable orbits. Yet, that he continued to worry about this problem seems to imply that he still thought in terms of many spiral projectiles extending beyond the South Pole.

In conclusion, no matter how much I try to give Newton the benefit of the doubt, he keeps bringing me back to those spirals.

Newton vs. Leibniz

Mentioning the three-body problem brings up one of the delightful double ironies in the history of science. Gottfried Leibniz, in Germany, was another of Newton's antagonists. They clashed on the nature of forces; Leibniz accused Newton of introducing occult powers back into science. Newton (wrongly) accused Leibniz of stealing the calculus from him. (Newton called his mathematical system fluxions.)

The development of Newtonian mechanics, along with the solution of the three-body problem—such that the solar system remains stable without the need of God's constant intervention—took place during the 18th century through the work of a series of Continental mathematicians (the Bernoulli family, Euler, D'Alembert, Lagrange, and Laplace) using and developing the calculus as formulated by Leibniz, because Newton's fluxions was much too cumbersome. So Newton's concept of force (repugnant to Leibniz) was filtered through Leibniz's calculus (offensive to Newton) and ultimately led to the version of Newtonian mechanics found in any textbook today—much to the chagrin, I suspect, of both of them.

10.4. Newton and Hooke: A Debate Over a Spiral

The following story is a sort of coda to this chapter: it also constitutes a transition to the next.

If need be, further support for my thesis (above) may be gleaned from the exchange between Newton and Hooke that transpired from November 24, 1679, to January 17, 1680. This important disputation, the various details of which I shall not be exploring here, has been exhaustedly studied by historians. I will, however, be relating this debate to the projectile motion issue, and for that I think I am the first to make that specific connection. I especially wish to emphasize regarding the ensuing debate that I am concerned neither with who is right nor with what historian's interpretation is correct. My focus is only directed toward what light this may or may not throw on the projectile motion topic.

Hooke initiated the exchange, which stimulated Newton to think deeply on the subject of celestial motion and especially of a celestial physics. Hooke stimulating Newton to think about science? What gives? Did Newton not inaugurate the topic with his celebrated thought experiment about an artificial satellite many years earlier?

Of course, and surely recalling the famous story of Newton and the apple is appropriate here. It is true that the insight into what became his mathematical theory of gravity came about when he realized that the same force acting on a falling body near Earth might also apply to the moon in its orbit. This event took place, Newton tells us, during the period (1665–1666) he spent on his mother's farm when Cambridge University was closed due to a plague. Whether the story is apocryphal or not, and even if it actually involved a falling apple (which Newton does not mentioned), we do not know. But clearly the famous projectile diagram embodies the essential insight of Newton. So if it came about in 1666, and I see no reason to doubt Newton on this, why then the significance of Hooke's rekindling the topic of a celestial physics? Well, for one, because Newton essentially had ignored it over those years. From the plague years to Hooke's letter, Newton devoted little effort to the topic; instead his prime intellectual effort was directed primarily toward theological and alchemical studies. (The next intellectual catalyst, as we know, was Halley's visit in 1684.) At most, Newton speculated on a possible cause of gravity, thinking in terms of an all-pervading ether filling space and providing the contingent mechanism for the pressure of gravity. Such an ether model was a common hypothesis in the 17th century, since it entailed what was viewed as a mechanical explanation based on direct contact rather than action-at-a-distance. The specific model Newton favored was borrowed, in part, from Descartes: the gravitational action around the sun, for example, arose from a vortex of ether swirling around it. Otherwise Newton's thoughts over more than a dozen years were on what we would consider non-scientific matters. Those insights during the plague had remained dormant.

Hooke's first letter (November 24, 1679) set forth two issues: introducing Newton to a hypothesis of his for celestial motion, and pursuing ways of proving the motion of Earth—both matters conceived of within the Copernican system. Hooke's hypothesis was a brilliant insight into celestial physics, using linear inertia and the realization that in circular motion the centrifugal force is caused by a tangential inertial power. Most historians are convinced that this is how Newton came to correct his previous erroneous assumption about rotational motion, and for this reason Hooke did more, much more, than just stimulate Newton's intellectual curiosity (see section 11.2). In his reply (November 28, 1679) Newton put forward a thought experiment ("a fansy of my own") directed toward proving the rotating Earth. We know that Galileo's insight into the relativity of motion had shown that an object dropped from a tower falls to the bottom despite the movement of Earth. As before, this result was true locally, where one need only approximate the result as if the motion were linear. But taking Earth as a whole, with its rotational motion, what would be the path of a falling body assuming a transparent Earth where the body falls to the center? Newton maintained that the falling body would first fall slightly to the east (since Earth rotates from west to east, and the tangential speed at the top of the tower is greater than at the bottom), after which, in passing through Earth, it would spiral toward the center. A spiral motion was compatible with his etheral vortex model.

Hooke replied (December 9, 1679) with an ingenious argument using Kepler's orbits, which clearly shows that the falling body attracted toward Earth's center would be constrained to follow an elliptical path. It seems that Newton recognized Hooke's insight, and perhaps thought he himself had made a mistake, for Newton did not defend his spiral in his last letter to Hooke (December 13, 1679). Newton, nevertheless, was not prone to accepting defeat, and in a terse reply he acknowledges that the path would not be a spiral, but argues that, in a rotating frame of reference, it would not be Hooke's exact ellipse either. The next two letters from Hooke (January 6 and 17, 1680), Newton did not reply to, at least not immediately, although a brief reference was made much later in the year in a letter to Hooke on December 3, 1680.

We find, significantly, in the ensuing writings of Newton, an ambivalence, or a hedging of bets, in his explanations of gravity. Is it an attractive force or an etheral pressure? The exchange with Hooke opened up these possibilities. Consider the following remarks taken from Newton's correspondence. Writing to Thomas Burnet, who had published a popular book on a hypothesis for the origin of Earth, Newton speaks of "ye pressure of ye vortex or of ye moon upon ye waters" (December 24, 1680). He repeats this twofold explanation in the next letter sometime in January, using the phrase "ye pressure of ye moon or vortex, etc. may promote ye irregularity of ye causes of hills." In short, either a vortex of ether or a gravitational power of the moon may explain the phenomenon. The next month (February 28, 1681), in a letter to the Astronomer Royal, John Flamsteed, on the comet of 1680 (see next chapter), he begins a sentence with, "The attraction of ye earth by its gravity," but a sentence later refers to "ye motion of ye vortex." This vacillation continued during the writing of the *Principia,* although with its publication there was a loss of confidence in the vortex model. Thus, at the end of Book II (on the motion of bodies in resisting media) he mathematically proves that Kepler's third law will not hold if a vortex around the sun is postulated, and he concludes that "the hypothesis of vortices can in no way be reconciled with astronomical phenomena." Significantly, in a letter to Halley during the writing of his treatise (July 27, 1686) he speaks of Hooke "correcting my Spiral"—an admission of a mistake which is not customarily part of Newton's demeanor.

Had Newton thus completely given up on the vortex model? Am I therefore in error for reading a spiral into Newton's projectile passages right through third edition of *Principia* published just before his death? Surely not, for we find in various published and unpublished writings after the first edition of the *Principia* Newton still speculating on sundry etheral models to explain gravity and other forces in nature. The most obvious are the ether conjectures found in his work on optics (see section 9.3), such as the queries added to the subsequent editions of the *Opticks* (see Chapter 9); moreover, over these same years he was revising the *Principia,* and still (I assert) hedging his bets in the projectile passage. [As the *Principia* went through three editions between 1687 and 1726, the *Opticks* went through five editions (two Latin and three English) between 1704 and 1721.] There is no reason to make a disjunction between Newton's thinking in the *Principia* and the *Opticks,* however much they differ methodologically. They both expressed his conceptual worldview.

Let me give the last remark to Newton. Over 30 years after the exchange with Hooke, Newton recalled the debate. As we have seen in a letter to Halley in 1686, he seemingly admitted to the spiral being a mistake, and hence conceded his error as being caught by Hooke. But the memory still smarted, it seems; it was not a mere spat over a spiral. For Newton, who never could get over being admonished by Hooke, is later quoted as dismissing the spiral as "a negligent stroke" of his pen!

Notes and References

Our argument is found in D. Topper and D.E. Vincent, "An Analysis of Newton's Projectile Diagram," *European Journal of Physics*, 20 (1999), pp. 59–66. The critique is Michael Nauenberg, "Comment on 'An Analysis of Newton's Projectile Diagram,'" *European Journal of Physics*, 21 (2000), pp. L5–L6. Our rejoinder is D. Topper and D.E. Vincent, "Reply to Comment on 'An Analysis of Newton's Projectile Diagram,'" *European Journal of Physics*, 21 (2000), pp. L7–L8.

Philip and Phylis Morrison, *The Ring of Truth: An Inquiry into How We Know What We Know* (New York: Random House, 1987), Newton's sketch is on p. 250.

Isaac Newton, *De Mundi Systemate (A Treatise on the System of the World)*, English trans. (London: Dawsons, 1969). The most recent translation (of the third edition) of the *Principia* is Isaac Newton, *The Principia: Mathematical Principles of Natural Philosophy*, I. Bernard Cohen and Anne Whitman, trans. (Berkeley and Los Angeles: University of California Press, 1999). I was able to reconstruct the history of various versions of the projectile passage using *Isaac Newton's "Philosophiae Naturalis Principia Mathematica,"* 2 vols., Alexandre Koyré and I. Bernard Cohen, eds. (Cambridge, MA: Harvard University Press, 1972).

On Henry Cavendish, see Christa Jungnickel and Russell McCormmach, *Cavendish: The Experimental Life* (Cranbury, NJ: Bucknell University Press, rev. ed., 1999), pp. 440–456, and B.E. Clotfelter, "The Cavendish Experiment as Cavendish Knew It," *American Journal of Physics*, 55 (March, 1987), pp. 210–213.

Three historical books that I found to contain the erroneous spiral: T.S. Kuhn, *The Copernican Revolution: Planetary Astronomy in the Development of Western Thought* (Cambridge, MA: Harvard University Press, 1985); S. Toulmin and J. Goodfield, *The Fabric of the Heavens: The Development of Astronomy and Dynamics* (New York: Harper Torchbook, 1961); and A. Koestler, *The Sleepwalkers: A History of Man's Changing Vision of the Universe* (Harmondsworth: Penguin, 1964).

Newton's thoughts on the stability problem are in his four letters to Richard Bentley (1692–3) reprinted in *Newton*, ed. I.B. Cohen and R.S. Westfall (New York: W.W. Norton, 1995), pp. 330–339; and Queries 28 and 31 of I. Newton, *Opticks* (New York: Dover Publications, 1952 reprint of the 4th English edition, 1730), pp. 362–370 and 375–406.

The correspondence with Halley and Hooke is in *The Correspondence of Isaac Newton, vol.2 (1676–1687)*, H.W. Turnbull, ed. (Cambridge: Cambridge University of Press, 1960).

The quotation dismissing the spiral as a slip of his pen, is in Richard S. Westfall, *Never at Rest: A Biography of Isaac Newton* (Cambridge: Cambridge University Press, 1980), p. 385.

11
A Change of Mind:
Newton and the Comet(s?)
of 1680 and 1681

I am holding a facsimile copy of the first edition of Isaac Newton's *Philosophiae Naturalis Principia Mathematica*. Published in 1687 it is commonly referred to simply as the *Principia*, as if there were no other "principles." The book, without a doubt, represents the culmination of the Scientific Revolution begun in the 16th century, when Copernicus proposed that the sun, not Earth, was at the center of the universe.

As I leaf through the book, I glance at numerous abstract geometrical diagrams with labeled triangles, circles, arcs of curves, and the like. Such illustrations would be expected in a book on the "mathematical [namely, geometrical] principles of natural philosophy [that is, science, in modern terminology]." After all, Newton's *Principia* was the foundation for what today is called Newtonian (or classical) mechanics.

Near the end of the book, between pages 496 and 497, I find a foldout illustration displaying Newton's drawing of the "great comet" of 1680–81 (Fig. 11.1). It is a picture, unlike the other abstract geometrical diagrams. It is doubly unique: it is the only foldout and the only actual picture in the entire book. When opened it measures 36 cm by 24.5 cm, twice the size of a page of the book. This important drawing appears near the end of the first edition of the *Principia* since the motion of comets was the conceptual culmination of the theory developed therein.

The image drawn by Newton depicts the comet as he observed it from November 1680, when it first became visible, into March 1681. In late November it had disappeared in the glare of the sun only to reappear in mid-December. Of course, during this time it swept around the sun, as is indeed drawn by Newton. But in the autumn and winter of 1680–1, while observing the comet, he did not make this deduction. In fact, he thought the comet that appeared in December was a different one from that which had disappeared in November. It was not obvious to him, nor to his friend Edmund Halley, that the comets were one and the same, because, in part, both men thought comets traveled in straight lines. (Incidentally, this was not "Halley's comet"; that one came the following year, 1682.) This chapter is about the transition in Newton's mind from the two- to the one-comet theory.

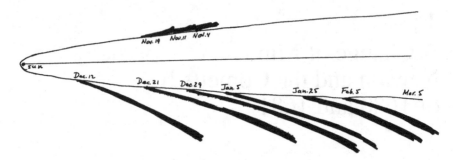

FIGURE 11.1. Newton: diagram of comet. A sketch of Newton's drawing of the great comet of 1680–1681, displayed as a fold-out in the first edition of the *Principia*. Note how the diagram shows the fact that the comet's tail increases on it return trip from the sun.

11.1. Comets and a Celestial Physics

Newton's conviction that there were really two comets had ancient roots. Aristotle placed comets below the moon, because they were assumed to be unpredictable, transient, and ephemeral entities. Only the heavens were perfect, permanent, and unchanging. As well, within the astrological tradition, comets were seen as omens of disasters, and the modern separation of astrology from astronomy was only taking place during Newton's life.

The late 16th century saw the transition toward the modern theory of comets. Tycho Brahe, in attempting to measure the parallax of several comets (starting with the great comet of 1577), concluded that they were beyond the moon, within the celestial world of the sun and planets. It was also noted among astronomers that the tails of comets always point away from the sun; this comet–sun connection reinforced the celestial placement.

At the time it was still assumed that planets were attached to celestial spheres, which provided the mechanism for planetary motions. Tycho concluded from the apparent paths of the comets, however, that celestial spheres did not exist, for the comets seemed to pass right through them. This was ostensibly a reasonable conclusion and certainly would be convincing today. Yet it was not the only possible explanation at the time. Just as it was not obvious that comets orbited the sun, it was not obvious that they were material entities. If, say, they were merely made of a light-like substance, then, being essentially a beam of light, comets could pass right through celestial spheres in any direction. One hypothesis of a comet's tail: it was light refracted through a translucent body (visualize the reflection of the setting sun over a body of water). Today we know that tails are the result of the evaporation of ice heated by the sun and swept away by the solar wind; hence the tails always point away from the sun. This heating increases as the comet rounds the sun producing a longer tail during recession, as is indeed illustrated in Newton's diagram, even if he did not realize the significance. Rather, it seems the differences in appearance supported there being two different comets.

Newton accepted the celestial abode for comets but initially believed they were transient bodies, betraying a remnant of Aristotle's cosmology. It is therefore not unexpected that he thought the comet of November 1680 was on a linear path, and the "new" comet of December 1680 to March 1681 was another one on a different straight path. Both Kepler and Galileo held to the linear path of comets.

But, as seen in the foldout, Newton had a change of mind, at least by the time he published the *Principia*. When and why?

In the spring of 1681 John Flamsteed, the first Astronomer Royal at Greenwich Observatory, put forward the notion that the two comets were one and the same and that its path was bent by the sun. A source of his supposition was his belief that the sun possessed a magnetic power that acted on the comet, first attracting it toward the sun, and then repelling it away. The idea of magnetic powers in the heavens was common at the time, ever since another Englishman, William Gilbert in 1600, had discovered that Earth is a magnet. (On Kepler's speculations, see below.) Newton rejected Flamsteed's mechanism, however. Since Gilbert reported that magnetized iron loses its power after being placed in a red-hot fire, Newton was sure the heat from the sun would destroy any magnetic powers possessed by comets. A letter exchange between Newton and Flamsteed, however, did get Newton thinking about the need for some sort of mechanism *if* the two comets were really one.

Importantly, at this time Newton was also working on the problem of a mechanism for the motion of the planets around the sun. Today the mechanism is simple and obvious, but it was not so then. The best scientific minds were pondering this problem, and the solution emerged from several false starts. Let's review the germinal thinkers. Kepler was one of the first to conceive of a "celestial physics"; the term was used in the subtitle of his 1609 *Astronomia Nova (The New Astronomy)*, a book containing his discovery of the elliptical orbits of the planets and the law that the planets sweep out equal areas in equal times. This was probably the most radical departure from ancient astronomy since Copernicus switched Earth and the sun. Inherent heavenly circular motion was taken for granted since ancient times, witness Galileo's circular obsession (see Chapter 8).

Perhaps almost as significant as Kepler's break from the hegemony of circular celestial motion was his concept of a celestial physics. To see why, you must steep yourself in the ancient cosmos, especially as conceived by Aristotle, where the division between the terrestrial world (below the moon) and the celestial world (beyond) was absolute. These two "worlds" were not only composed of different substances (earth, water, air, and fire down here, and ether filling up the heavens) but entirely different laws governed their motions. Circular motion was natural in the heavens, whereas natural physical motion on Earth was either toward Earth (gravity, from the Latin *gravitas*, or heavy) or away from it (levity, such as air rising in water). This meant that whatever physical laws controlled motion on Earth, they had no bearing on the motions of the planets. By speaking of a "celestial physics" (an oxymoron to Aristotelians), Kepler proposed a unity to the universe, a conception that allowed an earthly mechanism to be a model for celestial motion. Such an idea would have made no sense to Aristotle, or for that matter, almost any Aristotelian since antiquity (recall Galileo's demonstration with a ball in a bowl of water; see sections 4.4 and 8.4).

Kepler's celestial mechanism was based on Gilbert's discovery of Earth's magnetic power. Because of Kepler's homogeneous view of the Copernican universe, he thought that all the planets therefore must possess such powers (Earth was not unique under his interpretation of the Copernican system; after all, it was another planet). Hence there were magnetic attractions and repulsions directly between the sun and each planet. These powers caused the planets to move closer and further from the sun as they orbited in their elliptical paths, although Kepler simplified it to merely an eccentric one. To make this model work, Kepler used the fact of Earth's tilt; thus one magnetic pole predominates with respect to the sun. He then was forced to assume that the sun's magnetism was unipolar, not bipolar (see Fig. 11.2). As the planet orbits the sun, either the north–south interaction between them predominates, resulting in an attraction, or a north–north interaction leads to repulsion. In addition, a power (transverse to the magnetic push and pull) is needed to move the planets around in their orbits; this power, which he called the *anima motrix* (vital motion), emanated from the sun rather like spokes from a wheel, as (he hypothesized) the sun rotated. (Galileo later provided evidence for this when he saw sunspots move across the face of the sun; Kepler was delighted to find that the sun indeed rotated, as he predicted.) The resulting motion of each planet was therefore due to a combination of these

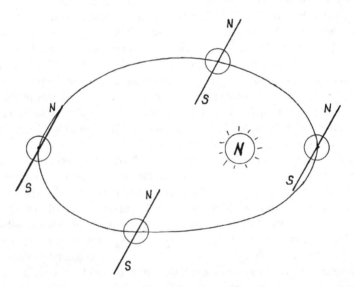

FIGURE 11.2. Kepler: celestial physics conception. Based on the discovery of the magnetic power of Earth, Kepler postulated magnetic powers between the sun and all the planets, and accordingly tried to account for the eccentric nature of their orbits. Since Earth is tilted in its orbit, it acts as a magnet with respect to the sun with either N or S predominating. To make the model work, however, it is necessary to assume the sun's magnetism is a monopole.

two powers: the *anima motrix* sweeping the planet around the sun, and the magnetic force pushing and pulling it into its eccentric orbit.

Kepler also coined the word *inertia* to describe the need for the constant *anima motrix*. This was partially right, since bodies at rest do tend to stay at rest. It was Galileo who added the next piece, conceiving of inertia as not only the tendency of a body to stay at rest but also the tendency of a body once set in motion to stay in motion (in a vacuum), without the need for a constant force. He then applied this to the vacuum of the heavens, arguing that the planets, once given the required push, would move in perpetual circles. Thus although Galileo has corrected Kepler's conceptual error on inertia, he erroneously deduced that the resulting motion would be circular. Moreover, he ignored Kepler's discovery of elliptical orbits. Such was the robustness of the ancient circles; to his dying day Galileo stubbornly held to circular motion for the planets (see Chapter 8).

This was straightened out by the Frenchman, René Descartes, who presented the first clear (modern) definition of linear inertia in his *Principia Philosophiae*. (Note how the full title of Newton's *Principia* was a variation of Descartes's work, as if Newton were writing the *mathematical* formulation that Descartes could not achieve. In the end, as noted, Newton's *Principia* ultimately became *the* Principles.) Descartes's *Principia* was published in 1644, and 20 year later the young Newton scrutinized it in detail, while reading works that were not on the curriculum at Trinity College. It most probably was his source of the concept of linear inertia. The weighty problem was how to work linear inertia into a model of a celestial physics.

I must confess that my previous description of Kepler's celestial physics is a bit of a caricature. The essence is true: he postulates both a rotational power sweeping the planets along and an independent magnetic power attracting and repelling them out of a circular orbit centered on the sun; moreover, his source of magnetism is Gilbert's discovery of the magnetism of Earth, which Kepler notes as being directed along a north–south axis of Earth, and from this he infers the magnetic power of all the planets and the sun. Yet the entire discussion is presented in a cryptic, convoluted, and often dense exposition entailing what most would call today physics mixed with fantasy, astrology, and theology. One problem, for example, clearly arises from his diagram of celestial physics (Fig. 11.3) as presented in his *Epitome of Copernican Astronomy* (1618–1621). Using arrows for the magnetic axes, he draws them in the plane of the planet's orbit, as if he is merely flattening my diagram (Fig. 11.2) for simplicity. But nowhere does he explicitly relate the direction of the arrow to Earth's axis (or those of the other planets); that is, he does not explicitly connect the tilting of Earth to the magnetic orientation in his diagram. Worse still, the directions of Kepler's arrows in the plane are wrong; if we flatten Figure 11.2, the arrows should point in the direction along the perihelion-aphelion axis. His arrows, however, are perpendicular to that axis, from which he still infers the noncircular orbit. It seems to me, on the contrary, that magnets in this direction will not result in the planet coming closest at the perihelion and furthest at the aphelion, while the *anima motrix* sweeps the planet along. It only works for the arrangement in Fig. 11.2.

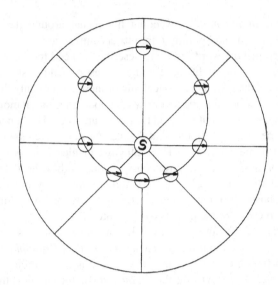

FIGURE 11.3. Kepler: celestial physics diagram. Fig. 11.2 is a conceptualization of Kepler's idea of a celestial physics. This sketch, however, is Kepler's diagram of the model, and the arrows, strangely, pointing in the wrong direction. From his *Epitome of Copernican Astronomy* (1618–1621).

All this confusion cannot be attributed to a lack of knowledge by Kepler. He knows about the tilt of Earth, its orientation in the plane of the orbit, and the precession of the equinoxes. Indeed, I wish to close this matter by relating it to an extremely interesting footnote in the second edition (1621) of the *Mysterium Cosmographicum*. The section in question involves Copernicus adding the conical "third motion" to Earth to account for its fixed orientation over a year (such that, in Kepler's words, "It always slopes towards the same point in the universe"), except for the small slippage to explain the precession of the equinoxes (see section 3.5). In the first edition (1596) he restates Copernicus's argument, since Kepler too is still thinking within the celestial spheres model. After presenting the basic argument he concludes by pointing to the consequential elimination of the (extra) conical motion of the (extra) celestial sphere—"that vast, monstrous, starless [and unnecessary] ninth sphere." Twenty-five years later, in the second edition, he adds a footnote to this section. Since writing the first edition Kepler has recognized Tycho's rejection of celestial spheres. As he puts it, "Tycho Brahe rightly accused me of this ancient and erroneous belief about the solidity of the spheres, and when he had read my little book wrote to me on this topic." Without the spheres, the third motion is eliminated, but what then supports Earth's tilt "towards the same point in the universe"? Here, I assert, Kepler comes close to Galileo's concept of inertia as a cause. It is true that when Kepler introduced the term *inertia*, his conception was not far from Aristotle's, namely, to characterize a body's resistance to being moved and retaining the need for a constant power to keep the body moving (hence the *anima motrix*). Thus when Kepler posits a power in Earth to maintain its orientation, he breaks the limits of Aristotle's conception, shifting toward Galileo's idea of the body also resisting a change in orientation. (Again recall Galileo's experiment with the ball in the bowl of water.)

Nevertheless, and not surprisingly, it is not this simple. We are dealing with Kepler! As I read the passage quoted below, Kepler gives three reasons for the fixed slope of Earth: inertia, magnetism, and rotation. That is, in addition to the natural (inertial) power of matter to remain in place (and hence Earth's fixed tilt), there is something about magnetism itself that keeps it in its place, and furthermore the daily rotation of Earth entails mechanical principles that have the same effect. He seems to be saying that all three are necessary to maintain Earth's tilt; or maybe all three work together toward that end, although I think he is just hedging his bets by putting forth all possibilities. Here is how he puts it:

While the Earth's globe travels round in its annual motion about the sun, all the time it keeps its axis of revolution always parallel to itself in its various positions, on account of the natural [inertial?] and magnetic tendency in its inner parts towards staying at rest, or even on account of the continuity of the diurnal rotation about its axis, which holds it upright, as happens with a top which has been set in motion and is spinning. Consequently just as this [fixed orientation] is not truly a motion, but is rather rest, similarly there is no need of an imaginary little [conical] circle [that is, the third motion].

This passage, especially the last sentence, I assert, borders upon Galileo's argument. Yet, of course, there is the faulty mixture of magnetism and rotation. Perhaps the most fascinating and ironic part of all this is the mention of a rotating top. Is it not ironic that he brings up the image of toy top to explain Earth's fixed position? He seems to believe that there are mechanical principles operating such that the rotating body maintains its given orientation. This is not only patently false, but as we know (see below), the motion of a rotating top is indeed conical and the source of Newton's explanation of the precession of the equinoxes. So here is Kepler bringing up the image of a toy top in the context of a discussion of Earth's tilt, which he knows has an ever-so-small conical motion. As a child did he not notice a top's conical motion? How close can one come to a discovery and still miss it?

A Celestial Toy: Newton and Precession

About 1652 the Dutch artist Rembrandt van Rijn produced one of his most poignant etchings of *Jesus Preaching*. At the feet of Jesus, and in the foreground of the picture, are two contrasting figures: an old man, with his eyes raised and a hand across his month, pondering the words of the young sage; and a little boy, lying on the ground, facing away from Jesus and foreshortened toward the viewer, who is drawing with his finger in the sandy soil, doubtlessly ignoring the words being preached. Next to the boy is his toy top—presently, but conceivably only recently, at rest—with the string following a serpentine path around the top and across the ground. The toy top is the closest object to the viewer in the pictorial space.

When Rembrandt was scratching this etching on a copper plate, across the channel Isaac Newton was about ten years old, a likely age for a boy to play with a top. It is a fantasy of mine that Newton not only played with tops but also pondered their motion—the curious conical motion. We do know, of course, that a conical motion (like that of a top) is entailed in the motion of Earth to account for the precession of the equinoxes, and that the physical laws underlying this motion have their origin in Newton's *Principia*. (I have found an instance where Newton explicitly mentions a toy top, using the phrase, "such as boys play with," but the context is optics, not mechanics; he tells of painting a top with "divers colours" and spinning it so as to mix them, thus producing a white or gray. See also section 9.4.)

Most textbooks discuss the motion of a top in the section on the conservation of angular momentum. A qualitative explanation, using the right-hand rule, twice, is this: if the top is spinning counterclockwise, mentally bend the fingers of your right hand around the top and hence in the direction of rotation; now make three right angles with your thumb, forefinger, and middle finger, aligning your thumb long the rotational axis and your forefinger downward (by gravity) toward the floor; your middle finger then points toward the torque moving the top counterclockwise in a conical motion. Hence the top precesses in the same (counterclockwise) direction that it rotates. Now applying the same principles to the rotating Earth: the rotational axis still points "up," namely north, but the gravitational force is from the moon and the sun, trying to aligning Earth's axis perpendicular to their planes of orbit (rather than "falling," as the top) and thus the torque points in the opposite direction, propelling Earth in a clockwise conical motion. So Earth precesses in the opposite direction that it rotates (Fig. 11.4).

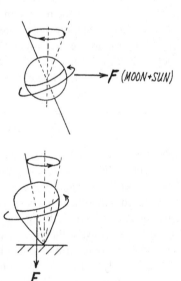

FIGURE 11.4. Newton: precession of a top and Earth. A diagram illustrating that the forces used to explain the conical motion of a toy top are similar to those acting on the spinning Earth, and thus providing a mechanical explanation of Earth's precessional motion.

Newton appropriately is given credit for first working this out. But his results were only rudimentary; his attempt at a detailed calculation was doomed to failure because he did not have the concept of angular momentum, and he was not able to deal with rotational motion of rigid bodies. According to Richard Westfall, Newton "doctored" the calculations "in his effort to create an illusion of great accuracy." In short, Newton's original mechanics was nowhere nearly as rigorous as that presented in modern textbooks of Newtonian mechanics.

Nevertheless, in light of the separation of the terrestrial and celestial worlds since ancient times as well as the gradual embracing of mechanical models for both earthly and celestial phenomena in the 17th century, I think this story of the *conception* (even if the execution was faulty) of Earth as a spinning top—the mental image alone is marvelous—is one of the most extraordinary case studies in the rise of a viable celestial mechanics.

11.2. Newton's Struggle with a Celestial Physics

One of Newton's early attempts at a celestial physics was based on the interaction between a centrifugal force outward and a centripetal force inward. Take a weight attached to a rope and spin it in a circle; there is a centrifugal force outward, as the spinning weight pulls on the rope. The very idea that such an "earthly" motion could be applied to the planets at once embodies Kepler's notion of a celestial physics and was anathema to Aristotelian physics. Newton was looking for a celestial physics based on the mechanism of the weight on a spinning rope, but he was not sure how this would work, since he thought of inertia in this system as only pulling away *radially* in a straight line from the center. Therefore, there was a balance between the centrifugal force (outward) and the centripetal pull (inward) by the rope.

The Dutchman Christiaan Huygens coined the term *centrifugal*. Newton first used the word *centripetal* for the force inward (by the rope) in his little treatise *De Motu* (On Motion; see section 10.1). Both men independently deduced the important mathematical formulation that for such a circular rotation, these forces are proportional to the speed squared divided by the radius. From this Newton probably first deduced the inverse-square law.

Assume a body is moving in a circle of radius r with speed s, and making one revolution during a period of time T. The distance traveled in T is the circumference $2\pi r$. Using the above-mentioned formula ($F \propto s^2/r$), results in $F \propto 4\pi^2 r^2/T^2 r$, which reduces to, $F \propto r/T^2$, since constants can be eliminated from this proportion. Here Newton applies Kepler's third law ($T^2 \propto r^3$). The result is

$$F \propto 1/r^2.$$

This is the famous inverse-square law, which later formed the mathematical foundation of his theory of gravity.

When Newton actually first derived this law is a matter of dispute among historians, the earliest being the so-called "plague years" (1665–6) when Cambridge University was closed and Newton was home on his mother's farm. It is important to note that this result is only an approximation for a circle. To derive this for an ellipse was a formidable mathematical problem (see section 10.1). Most significant is that Newton here applies a planetary law (Kepler's) to a mechanical (terrestrial) problem. This reveals further that he was thinking within the framework of a celestial physics.

Returning to the "balance" between centrifugal and centripetal forces, there is a common misconception about such motion. Ask a novice what happens if the rope is released, and the usual answer is that the weight will fly off in a (radial) line in the direction of the centrifugal force. Such was Newton's first model applying linear inertia to rotational motion. But it is wrong; when released, the weight actually flies off *tangent* to the circle, since a tangential linear inertia is the source of both the circular motion and the centrifugal force. The centrifugal force is really caused by an *imbalance* between inertia and the pull of the rope. Initially, in thinking about mechanical motion applied to the heavens, Newton was thinking like a novice (Galileo grappled with this too; recall section 8.3).

It is no wonder that when confronted with the comet problem of 1680–1 Newton did not leap to the conclusion that the two comets were really one comet orbiting the sun, since as yet he had no clear mechanism even for the planets. It was not apparent to him how to deal with the powers and forces acting on the planets as they move around the sun.

Yet, as a matter of fact, Robert Hooke provided the answer a year before, during the significant exchange of letters with Newton in the winter of 1679–80 (see section 10.4). Hooke realized that a celestial physics could be simplified by combining linear inertia as a power tangent to the orbiting body's path at each point with an attractive power (later Newton's centripetal force) toward the sun. Hooke's insight conceptualizes a body's path as "that of compounding the celestiall motions of the planetts [by means of] a direct [inertial] motion by the tangent and an attractive motion [or force] towards the centrall body." Centrifugal force consequently ensues (seemingly created by the motion) as the tendency of the planet to depart continually from its linear path, with inertia being the power "sweeping" it along. Simple and brilliant—a mechanism found in any textbook today—but, and here's that mantra again, only obvious in retrospect.

Hooke's gift came with a bonus. He brilliantly showed that such a mechanism would also work for an elliptical (or at least an oval) motion, using the example of a pendulum—another earthly mechanical device (like a spinning top). He begins with the observation that an oscillating pendulum bob suspended from a wire, and given a slight lateral motion, moves in an oval. Analyzing the forces on this pendulum reveals that the oval motion alone is due to a force directed toward the center.

Here's how it works. Looking at the oval motion of a pendulum, two components of force (g and F) act on the bob (Fig. 11.5). Force g is the gravitational force; force F is on the wire. But F can be decomposed into e and f, and since e is balanced by g, then f alone provides the force inward. Hence force f (directed toward the center of the oval) and inertia (tangent to the oval) are sufficient components that explain the oval motion of the bob. This, Hooke argues, is analogous to the elliptical motions of the planets.

Hooke, and eventually Newton, became convinced of the truth of Kepler's elliptical orbits. Hence arose the possibility that the elliptical motions of the planets could be caused solely by a pressure or force directed toward the center of motion, along with inertia.

An aside: later, when Newton came to write his *Principia*, he showed that Hooke's physical intuition could be subordinated to mathematical analysis; he specified the magnitude of the centripetal force and demonstrated that for an elliptical motion this force would be directed not toward the center of the ellipse, as Hooke thought, but to one of the foci, where Kepler placed the sun.

Strangely, in the late autumn and winter of 1680–1, a year after this vital exchange with Hooke, Newton did not apply this solution of a working model for a celestial physics to the problem of the comet (or comets). We find him still waffling about the mechanism. This is revealed in letters (and drafts of letters)

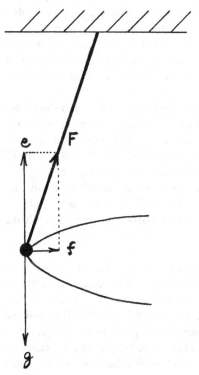

FIGURE 11.5. Hooke: elliptical motion. The force of gravity (g) and the tension (F) in the wire act on the pendulum as it moves in an oval plane. The force (F) can be reduced to its two components: e and f. Since $e = g$, and f is directed toward the center of the oval, then Hooke inferred that the elliptical paths of the planets could be explained by one central force (including inertia).

sent to Flamsteed in the spring of 1681. Recall, Flamsteed proposed that there was only one comet and its motion was based on a magnetic power between it and the sun. Newton remained more committed to the two-comet theory but he did entertain the possibility of there being just one comet and mentioned possible mechanisms. He writes that the comet's motion around the sun may be due to a compounding of the centrifugal and centripetal forces—the erroneous model. Perhaps even more bizarre from a modern viewpoint, he speculates that the comet may turn around across the *front* of the sun (not behind it, as we know today), due apparently to some repulsive force between the comet and the sun. Obviously Newton was still grappling with what was for him a staggering problem.

It was not until the autumn of 1684, about four years later, in what became a draft of the *Principia* (the little treatise, *De Motu*) that we can document Newton using the correct model for the planets (see section 10.1). Unfortunately, we do not know what moved him, but clearly something happened to persuade him that Hooke's solution might be correct. Significantly for this story, he also attempted to compute the orbit of comets, as if they too (like the planets) travel around the sun and move in ellipses. It was a formidable mathematical problem and the result was incomplete. Consequently, he was still not convinced of the one-comet hypothesis. This can be seen from a remarkable letter to Flamsteed in September 1685, written while he was working on the section of the *Principia* on comets. He says he is in the process of calculating the orbits of comets, particularly that of 1680–1, because, "it seems very probable that those of November and December were ye same comet." It is almost farcical to picture Newton, writing what we know will be his masterpiece, the culmination of which was the application of his model to comets (one third of the third section was devoted to the topic in the first edition), and realize he *still* is waffling a bit, yet conceding that the one-comet idea is "very probable."

Almost amusingly, too, we find that in the original copy of the letter the phrase quoted above is underlined twice (apparently by Flamsteed, upon receipt of the letter), who also wrote in the margin: "He [Newton] would not grant it [namely, that Flamsteed was correct] before [,] see his [Newton's] letter of 1681 [above]." Flamsteed was gratified; Newton was finally coming around to his view.

Indeed he did. Comets make their appearance at the end of the *Principia*, where Newton applies his mathematical skills to their orbits and proclaims their elliptical forms. Specifically the comet of 1680–1 is studied, with the accompanying picture (Fig. 11.1).

To flesh out this story, here are Newton's comments on the comet of 1680–1 in the various versions of the *Principia*. In the draft of the popular account of the theory, which was never published in his life (but posthumously titled, *A Treatise on the System of the World*; see section 10.1), he conceded that "hitherto [I] considered those comets as two, yet, from the coincidence of their perihelions and agreement of their velocities, it is probable that in effect they were but one and the same." In the manuscript to, and in the first edition of, the *Principia* he spoke of one comet around the sun. In the second edition (1713) he wrote, "It is abundantly proven that it was one and the same comet that descended towards the sun in November as ascended from it in the following months." Curiously, he crossed out this sentence in his annotated copy of the second edition. Finally, in the third and last edition, written about a year before his death in 1727, he reiterated that "it was one and the same comet that appeared in the whole time from the 4th of November [1680] to the 9th of March [1681]." What first he thought were two comets turned out to be really "one and the same." And of course that was how it was depicted in the foldout picture of the one great comet of 1680–1, drawn by Newton himself.

Newton and the Theology of Comets

The modern physics of comets, as seen, is indebted to the genius of Newton. Moreover, during Newton's time other aspects of what ultimately constitutes modern science emerged: the separation of alchemy from chemistry, and astrology from astronomy, with alchemy and astrology eventually relegated to the realm of pseudoscience. In addition, comets were no longer portents of disaster, at least in the folk sense of predicting plagues, wars, the deaths of kings and queens, and the like.

Nevertheless, a continuing fascination with comets arose amid a flurry of hypotheses about comets as the cause of historical cataclysms, especially biblical events such as the plagues of Egypt and the Deluge. Newton knew of these ideas and took them seriously, very seriously. In this context, comets remained emissaries of disaster, not unlike the present-day argument about the extinction of the dinosaurs and other species around 65 million years ago caused by either a comet or an asteroid.

Having worked out that comets, like planets, may orbit the sun, Newton was confronted with a potential problem. Since their orbits were not within the ecliptic (namely, the plane of planets), the physics of motion alone entailed the possibility of a collision of a comet with Earth. For Newton this was a theological problem. How could a benevolent God permit such a malevolent deed? Such questions were more important to Newton than mere scientific puzzles. In fact, most of Newton's intellectual life was spent pondering issues anathema to today's science, such as matters of alchemy, biblical chronology, and church history. Actually only a very small part of his life's work was spent on science. As in the Middle Ages, theology was still the queen of the sciences for Newton.

So at length Newton produced a theology of comets too, based on his conviction of a benevolent Creator. Comets are actually divine agents, their "exhalations" restoring the solar system of needed substances otherwise being continually exhausted, such as water for Earth and fuel to the sun. Adopting modern jargon, this hypothesis may be seen as supporting a steady-state model of the solar system, one that does not run down or change over time, with Providence acting through comets to resupply the needed materials. Such an idea may also bring to mind the present-day notion of matter being created out of the vacuum of space, which incidentally was part of the steady-state model (see section 5.4).

One more point on the comet of 1680–1. Newton believed that this comet almost grazed the sun and that it would eventually crash into it. (Note how he drew its perihelion close to the sun.) A direct hit would over-fuel the sun, producing a conflagration (indeed, the final conflagration!) that would destroy Earth. He calculated the period of this comet as 575 years, and thus the "end of the world" was predicted as some multiple of $1680 + 575\ x$, where x is an integer. Thus for $x = 1$, the end is 2255. I have already seen some reports in the popular media on this prediction, coming with the authority of Newton, no less. In fact, however, the great comet of 1680 probably will never return; its period is extraordinarily long, so long that astronomers cannot predict its return. Incidentally, and finally, Halley's comet of 1682, which of course does return, is the only naked-eye comet with a rather short period.

Notes and References

For more on Newton and comets, see Simon Schaffer, "Newton's Comets and the Transformation of Astrology," in *Astrology and Science and Society: Historical Essays*, Patrick Curry, ed. (Woodbridge, Suffolk, UK: Boydell Press, 1987), pp. 219–243; and Sara J. Schechner, *Comets, Popular Culture, and the Birth of Modern Cosmology* (Princeton: Princeton University Press, 1997), especially chapters 6 and 7. For information on the 1680–1 comet and Newton's calculation of the end of the world, I am grateful to Sara Schechner (via the Internet).

Isaac Newton, *De Mundi Systemate (A Treatise on the System of the World)*, English trans. (London: Dawsons, 1969). The most recent translation of the *Principia* is Isaac Newton, *The Principia: Mathematical Principles of Natural Philosophy*, I. Bernard Cohen and Anne Whitman, trans. (Berkeley and Los Angeles: University of California Press, 1999). For a scholarly version, with delineations among the three editions, see *Isaac Newton's "Philosophiae Naturalis Principia Mathematica,"* 2 vols, I. Bernard Cohen and Alexandre Koyré, eds. (Cambridge, MA: Harvard University Press, 1972). Indispensable is *The Correspondence of Isaac Newton, vol. 2 (1676–1687)*, H.W. Turnbull, ed. (Cambridge: Cambridge University of Press, 1960).

Kepler's *Mysterium Cosmographicum* has been translated as *The Secret of the Universe*, by A.M. Duncan (New York: Abaris Books, 1981), material quoted from pp. 83 and 91.

My argument on Kepler's celestial magnetism is support by passages in his *Epitome of Copernican Astronomy* (1618–1621). I have used the translation of sections of this work from volume 16 of the *Great Books of the Western World* (Chicago: Encyclopaedia Britannica, 1939). See especially pp. 899 and 935–940.

For her invaluable assistance with Latin translations here and throughout this book, I am grateful to Professor Jan McTavish, Alcorn State University, Mississippi.

12
A Well-Nigh Discovery:
Einstein and the Expanding Universe

Einstein's stubbornness about his theories and his attitude toward experiments were explored in Chapter 1. Here I return to Einstein, with the focal point being his obstinacy over his cosmological model. This personal drama played itself out in the context of the cosmological revolution of the 20th century—a change, I shall show, that was as extensive and momentous as that initiated by Copernicus in the 16th century.

12.1. Nebulae and Galaxies

On a clear night, far from city lights, the sky is filled with thousands of stars, the moon may be seen in one of its phases, one or more of five planets are often visible, the Milky Way may stretch across the heavens, and, if one looks very, very closely, a few little blurs or smudges may be barely seen among the stars. The ancients bestowed the term *nebula* (plural *nebulae*) upon these several indistinct objects, from which is derived the English word *nebulous*. In the Northern Hemisphere two prominent nebulae are visible in the constellations of Orion and Andromeda; that they always remain within each constellation indicates that they are fixed in the heavens, rather like the stars (unlike, say, comets or meteors), yet their indistinct (non-starry, non-twinkling) nature puzzled the ancients. Since only a few existed, little effort, therefore, was directed toward understanding them.

The history of these little blurs occupies a most extraordinary story, for they are at the center of nothing short of a cosmological revolution that transpired in the last century. Some of these blurs subsequently were found to be entire galaxies; that revelation led to a rearrangement of the universe and the relocation of galaxies from being remote celestial objects to being fundamental cosmological entities, center stage as they replaced stars as a key structural component of the universe. By the 1930s they were the buildings, the stars were the bricks, and, indeed, clusters of galaxies were the neighborhoods in the landscape of the cosmos as it came to be understood by the 20th century and bequeathed to us.

An outline of that story begins in the 17th century with the invention of the telescope and the awareness of two things: that there are more nebulae than the few visible with the naked eye, and that the Milky Way is not as it appears—it is actually a dense collection of stars, not a continuous cloudy mass (see section 4.5). Over the following two centuries, the number of nebulae discovered increased from hundreds to thousands. Obviously, speculation about their nature accompanied this work. Since some nebulae were shown to be dense concentrations of stars (as was the Milky Way) initially a number of astronomers assumed that all nebulae were merely star clusters. Yet many nebulae stubbornly continued to appear blurred, even as telescopes were improved, implying that they were probably just masses of hot gases.

A most important astronomer in this story was William Herschel, most famous for having discovered the first planet (Uranus) not visible to the naked eye. Herschel devoted much of his observational time to the nebulae; he discerned various odd shapes among the nebulae, as he and his sister Caroline catalogued about 2500 of them. He made another discovery that later would prove important to this story—that the Milky Way is not actually a massive concentration of stars "out there" but that we (namely, our solar system of the sun and planets) are *within* it. Herschel, moreover, asserted that we were near the center of this massive star cluster. Copernicus may have shifted the positions of Earth and the sun, but Herschel centralized our sun-system within the Milky Way.

Since the Greek term for milky-like is *galaxias*, the Milky Way was often also called the galaxy. With Herschel's discovery, some speculation arose about the possible existence of other such galaxies beyond ours, and not surprisingly likely candidates were the nebulae—that perhaps nebulae were other galaxies, clustering masses of numerous stars. The term *island universe* was coined for this notion that a nebula was a galaxy like ours. Most scientists, however, became more convinced that nearly all nebulae were composed only of hot gases and thus were not star clusters; at most, nebulae were proto–solar systems, rotating gases within our Milky Way. Further conformation of this came after the Herschels, especially by the mid-19th century when very large telescopes (in Ireland and England) revealed for the first time the *spiral* nature of many nebulae. Thus talk of island universes was then seen as mere speculation, more like science fiction than fact.

12.2. Einstein's Cosmological Model

From a theoretical point of view the cosmological revolution of the last century begins with the landmark paper by Albert Einstein in 1917. The idea for it grew out of his theory of general relativity, a theory of gravity, completed late in 1915 (see section 1.1). Just as relativity constituted a radical break with classical physics, Einstein's cosmological model was profoundly different from the Newtonian cosmos of the 17th century. (For convenience we call the latter Newtonian, just as we call the ancient one Aristotelian, even though elements of it were drawn from the ideas of Descartes and others throughout the Scientific Revolution.) In Newton's universe Euclidean space extends infinitely in all three dimensions (or, at least, "indefinitely," so as not to impose upon the territory of God, the only true infinite). Space was a passive receptacle for (inert) matter (possessing inertia) and forces (such as gravity), the latter filling space. Space, matter, and force were the trinity of Newton's universe.

Newton's cosmology, generally accepted by the mid-18th century, replaced the ancient cosmos, sustained throughout the Middle Ages and the Renaissance. The ancient model, expounded by Aristotle (among others) and supported later by philosophers and astronomers, was not only Earth-centered but also finite in size (volume), and encompassed by the sphere of the fixed stars—an impenetrable boundary comprising the edge of the universe. During the Scientific Revolution, this cosmos was altered in two fundamental ways: first, the sun switched positions with Earth as the central body and second, the center disappeared as the stellar sphere was shattered and the stars were envisaged as being scattered throughout space. The sun-centered solar system was hence somewhere or anywhere or nowhere (!) within this infinite and unbounded universe. The contrast with the finite and bounded cosmos was overwhelming: Newton's cosmos had obliterated Aristotle's.

Einstein's paper of 1917 is entitled, "Cosmological Considerations of General Relativity." In simple physical terms, here is what Einstein said. General relativity replaced the concept of gravitational force as an instantaneous action-at-a-distance with a field model, where gravitational force, like electromagnetism, is propagated at the finite speed of light. Moreover, the field emanating from matter is really a geometrical distortion of space by matter itself. Empty space alone, without matter, is Euclidean; the presence of matter distorts or warps space, and such a space is non-Euclidean. We do not directly see any of this because the distortion is into another (fourth) invisible dimension.

An analogy facilitates a mental picture: consider a two-dimensional person living in a flat (Euclidean) surface; a distortion of this space would not directly be seen by the flat person, but the distortion may be inferred by the behavior of objects moving in it; the behavior of moving objects entails an apparent force around the objects as they get caught up in the warping of the local space (see Fig. 1.2). For the flat person, the non-Euclidean warping of space into a third dimension is the cause of the illusionary force. Analogously, a concomitant deformation of space into a fourth dimension constitutes the gravitational field for us, as three-dimensional persons. Space is distorted by the presence of matter—apples, Earth, the sun, or stars—and this we perceive as gravitational force.

Having deduced this explanation for gravity (the fundamental component of general relativity), Einstein next asked this question: What happens to the entire universe if one sums up all the local distortions of space around matter? This summing up is essentially what Einstein did in the 1917 paper. Beginning with the equation of local gravity according to general relativity, he found that the summation of all local matter bending local spaces results in the warping of the totality of space into a finite yet unbounded universe.

Finite and unbounded, how can that be? Recalling the contrast between Aristotle's (finite and bounded) model and Newton's (infinite and unbounded) model, is this not a contradiction? Again, an analogy with our flat person's experience helps form a mental picture. Consider this person now living on a sphere: the space of this universe is surely finite (like Aristotle's universe and unlike Newton's), for the amount of space corresponds to the finite surface area of the sphere (the entire surface area being a function of its radius). Traveling in this space in a straight line the flat person would never come upon a boundary but would travel along the arc of a great circle, and hence the space experienced is unbound (like Newton's, and unlike Aristotle's). For us, therefore, as three-dimensional persons traveling in a straight line in a four-dimensional space, we would, in time (depending on the radius of our universe), return to the starting point, since the total space is finite. Thus, the universe as a whole forms an unbounded and finite non-Euclidean space.

There was, however, a glitch in Einstein's deduced equation of the universe. Besides being finite and unbounded, the universe seemed to be unstable. This was a strange deduction—so strange, in fact, that Einstein dismissed it as being impossible and immediately corrected the equation by adding a stabilizing term. The added term he called the "cosmological constant." Both the Aristotelian and Newtonian universes were static, since the notion that the universe as a whole is stable was an *a priori* postulate in all cosmological theories. So was Einstein's universe. He believed that his cosmological constant might be on par with other fundamental constants of nature, such as the speed of light.

Empty Space and Nothing

Since Aristotle's universe was of finite extent, a logical question arises: What is beyond the stellar sphere? It is a logical question from a modern point of view but it really makes no sense in ancient thought. The finite universe was all that there was; that was everything, so nothing was beyond the stars. Of course, I realize that in saying "nothing" I will not necessarily convey the real meaning, since today there is usually no semantic distinction between nothing and empty space. Thus saying that nothing is beyond the stars means, in the mind of many readers I should think, that the ancient universe was conceived of as a sphere floating in empty space. But that would be utterly wrong. "Nothing" is *not* empty space. Nothing means nonexistence. Anyone transported into "nothing" would not return. In contrast one may travel into empty space; as astronauts do (although in space suits with an oxygen supply), and they return. Hence, to conceptualize the ancient model one needs to keep in mind the essential distinction between empty space and nothing; indeed, the notion of empty space does not exist at all in Aristotle's science, since his finite universe was filled with air, water, earth, fire, and ether. At most there was the idea of a potential void but this was ultimately impossible, as air immediately rushes in to fill the void—hence, the medieval law that "nature abhors a vacuum."

Another way of conceptualizing this universe is realizing that the space of the spherical universe was not necessarily Euclidean (of course, neither was it non-Euclidean, in the modern sense); throughout the ancient, medieval, and Renaissance eras Euclid's space only existed in geometry texts. (Non-Euclidean geometry was invented in the 19th century by three mathematicians at about the same time, and although initially rejected as meaningless, it was accepted into the fold of conventional mathematics by the end of the century.) With the so-called breaking of the stellar sphere in the 17th century, and the corresponding idea of the entire universe extending far off in three dimensions, the space of this universe could only then be comprehended as a Euclidean space.

Einstein's use of non-Euclidean space in general relativity demonstrated that the distinction between nothing and empty space is not merely a trivial semantic one. In his theory, it is the geometrical property of space, specifically its (non-Euclidean) bending around matter, that constitutes gravity. So this space is not a passive receptacle for matter; instead, it is an active protagonist interacting with matter. Matter bends space, and the distortion results in matter appearing to have an attractive (gravitational) power. Thus there really is no "spooky" (see section 1.1) attractive force: Newton's trilogy of space, matter, and force is replaced by the dualism of matter and space, or really just energy (or mass-energy) and space, since $E = mc^2$.

Hence, a deep significance of Einstein's theory of general relativity is that space indisputably is not "nothing." On this, Aristotle was right.

12.3. Observational Astronomy in the Early 20th Century

Einstein, of course, was free to make abstruse calculations about the cosmos on the back of an envelope, but in the end it all had to jive with the real world of observational astronomy. We left that world (section 12.1, above) in the mid-19th century, with we humans comfortably at the center of our Milky Way and speculating about possible galaxies beyond. Maybe comfortably, but also trapped, because the first measurements of the distances of the stars, made possible with the improvements in telescopes, were made by a triangulation technique involving parallax (see Stellar Distances and Parallax, in Chapter 5). But by the late-19th century, astronomers realized that this method had a limit, due to number of causes (the orbit of Earth, the laws of optics for resolving light, and so forth). In short, the method can measure celestial distances only to around 500 light-years. Anything beyond that can not be measured, and importantly this empirical limit is well within the confines of our Milky Way. Hence, cosmologists (or poets) could speculate about the universe beyond our Milky Way but astronomers knew better. Our Milky Way really does embrace an "island universe" – at least, so thought the best astronomers at the term of the last century.

This parallax limit, however, was broken in the early part of the 20th century with a discovery by Henrietta Leavitt of the photometry department of the Harvard College Observatory, who studied a class of stars of regularly changing intensity. They were called Cepheid variables, since she first noticed them in the constellation Cepheus the king; subsequently they were found everywhere, yet the term stuck. The periods of fluctuation of these stars ranged from days to months. Another key discovery was made when several Cepheids were found within star clusters; by measuring their fluctuation periods it became clear that there was a correlation between the average intensities of the Cepheids and their periods, since all these clustered Cepheids were at the *same* relative distance from us. As she wrote, it is "worthy of notice that . . . brighter variables have the longer periods."

This important discovery was put to use by the astronomer Harlow Shapley at Mt. Wilson Observatory in California, who calibrated the scale of relative intensities to convert it into absolute values. This required knowledge of local Cepheids, that is, close enough to measure their distance by parallax; then, using the inverse-square law of the intensity of light, the absolute value of the Cepheids follows. (Here is an analogy if this is not immediately apparent: if you know the distance to a light bulb, then by measuring its observed intensity you can deduce its actual intensity, say 100 watts, since light intensity decreases inversely as the distance squared.) The result was a correlation between the periods and absolute intensities (or actual luminosities) of Cepheid variables, now known as the period-luminosity law (Fig. 12.1).

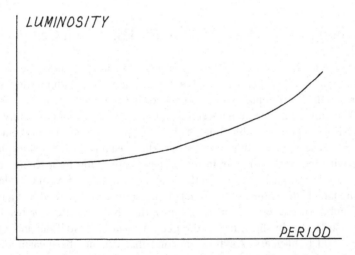

FIGURE 12.1. Period-luminosity law. A schematic diagram of Leavitt's law, showing the increase in the average luminosity of the Cepheid variables verses the periodic cycles of their change in brightness.

This law broke the 500 light-year limit. Here's how it works: observe a Cepheid variable and measure its period. From the period-luminosity law you obtain its absolute intensity or luminosity. Likewise the distance of the Cepheid is deduced. (Using the analogy again, you see a light at a distance and know its actual wattage, say 100 watts. Comparing this wattage with the observed intensity and using the inverse-square law you obtain the distance.) Thus, as long as we can see a Cepheid, we can measure its period, and hence obtain its distance. Cepheid variables became the first new signposts of the universe for the early 20th century. Importantly, Shapley used them to make the first real measurement of the size of our Milky Way and he discovered that it is about 100,000 light-years in diameter. (Actually, his initial calculation was too large by a factor of three, due to an erroneous calibration of the period-luminosity law.) But most significantly (and contrary to Herschel) he found that our solar system is really near the *edge*, not the center of the galaxy. By the 1920s this was the revised picture of our place in the Milky Way. But where were we in the universe?

Before pondering this question, there is another story in early 20th century observational astronomy that I have skipped—a discovery by the astronomer Vesto M. Slipher working at the Lowell Observatory in Arizona. Whereas Leavitt's discovery had immediate application, Slipher's was initially puzzling.

Starting in 1909, Slipher worked on spectroscopy of nebulae (essentially looking at the spectral colors of their light), believing they were proto–solar systems. He discovered that their light exhibited a shift toward either the red or blue ends of the spectrum. The first measurements were of the nebula in the constellation Andromeda, and he found the light shifting toward the blue. By around 1914 he had measured 15 nebulae, finding most shifting to the red; by 1917 he measured 25 nebulae and again most (22) were shifting toward the red. Today such red or blue shifts are obvious cases of things either receding or advancing from us, respectively; moreover, his discovery is seen as major step toward our knowledge of the expansion of the universe.

Looking at this from a pre-1920s viewpoint, however, Slipher's findings were more of a puzzle. First, the only known galaxy was still our Milky Way; true, there was speculation as far back as the 18th century that various nebulae may be other island universes, but this was more in the realm of science fiction. At the beginning of the 20th century most observational astronomers worth their salt were sure we were trapped in the Milky Way (within the 500 light-year limit), until Leavitt's discovery in the second decade. Second, although some nebulae were shown to be composed of stars (hence really star clusters, not just hot gases), nearly all still seemed to be just swirling masses of gases, and, as well, *within* our galaxy. Third, it was not widely accepted that the spectral shifts in light were caused by motion—namely Doppler shifts. Here is the historical background. (Note: some of the following discussion overlaps with, but also supplements, that in section 5.4.)

The Austrian physicist Christian Doppler in the mid-19th century discovered a shift in the wavelength of sound caused by the motion of the source. For an approaching wave, the wave is squeezed and hence the wavelength is shortened, and vice versa—a receding wave is stretched, lengthening the wavelength. (Since frequency is the inverse of wavelength, an approaching siren goes up in pitch and then goes down as it recedes.) By analogy then, light would shift toward the (shorter) blue as the source approaches and shift toward the (longer) red with recession. But there was no consensus among scientists on this analogy between light and sound. True, at the time they both entailed a wave model; however, light waves were transverse (think of a vibrating string) and sound waves were mainly longitudinal (think of a vibrating spring). Thus, many scientists did not interpret the red/blue shifts for light as Doppler shifts, and the mainly red shifts of the nebulae were a puzzle. Slipher first announced his discovery at the meeting of the American Astronomical Society (AAS) in August 1914; moreover, he interpreted the spectral shifts as Doppler shifts, but thought this implied that there was a relative motion of our galaxy through the universe. In the audience was a research assistant from Yerkes Observatory of the University of Chicago—Edwin Hubble.

Hubble finished his Ph.D. in 1917 and went to war. Returning in 1919, he landed a job at Mt. Wilson Observatory. His first few years overlapped with Shapley's last years there, for in 1921 Shapely moved to Harvard Observatory (see Shapley's Blunder, below). Despite the overlap, it seems they had little collaboration; apparently there was some tension between them. Instead, Hubble worked with his very able assistant, Milton Humason, using a recently constructed telescope, which was the largest in the world at the time. (Dominance in observational astronomy shifted from Great Britain and Europe to the United States in the 20th century, fueled largely by research funds from philanthropists.) In October 1923 Hubble found a Cepheid variable in the nebula in Andromeda. He later found more, and using the period-luminosity law (Fig. 12.1), he measured the nebula's distance, obtaining one million light-years (actually it is about $2^{1}/_{2}$ times that, the error due to a miscalibration of the period-luminosity law again). The result was astounding: there was little doubt that this nebula was *external* to our galaxy. Our isolation within our island universe was no more; for the first time there was empirical evidence, not just poetic speculation, that something exists outside the Milky Way—and so possibly did other nebulae. In January 1925 Hubble announced the discovery at the AAS meeting. In the following years Hubble and Humason found other external nebulae.

But this was only the beginning. Hubble and Humason also commenced a program built upon Slipher's work, by measuring the spectral shifts of dozens of external nebulae. Most exhibited a red shift in their spectra. And they found something else: for nebulae close enough to measure their distance, a correlation was found between the *amount* of red shift and the distance. Plotting these variables, Hubble found a linear relationship, which he first published in 1929 using data from 46 nebulae; later, with data from 40 more nebulae, he further confirmed the linear law in a second paper of 1931 (Fig. 12.2). How were these data to be interpreted? Hubble was cautious. One possibility was that the red shifts were Doppler shifts and hence all these galaxies were receding from us, with only a

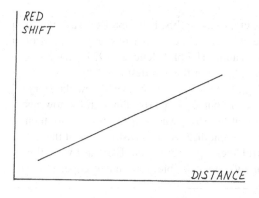

FIGURE 12.2. Hubble's law. A schematic diagram of Hubble's discovery of the linear correlation between the red shift of galaxies and their distances.

few (e.g., the galaxy in Andromeda) advancing. Some scientist jumped to this conclusion; if so, it seemed to imply that the universe is expanding, as some theorists (as we shall see below) had proposed. Others suggested different causes for the red shifts: that light slows down by the gravitational pull of the stars and nebulae as it travels and bends through space, the so-called tired-light hypothesis (see section 5.4).

While Hubble was pondering these matters at Mt. Wilson, he met Einstein, who, in the winter of 1930–31, was on the first of what became three annual sojourns to California. How did their meeting go? Before looking into this we need to go back to where we left off in theoretical cosmology.

Shapley's Blunder

The following incredible but undocumented story has circulated among astronomers for decades.

When Hubble discovered a Cepheid variable in the Andromeda nebula, he wrote to Shapley, who was now at Harvard, announcing his discovery. (Shapley had landed the directorship position of the Harvard Observatory, where Henrietta Leavitt worked. When he moved there in April 1921, sadly she was dying of cancer. Unfortunately, she could not work very much, so there was little collaboration, but he was with her at her death. In his autobiography he calls her "one of the most important women ever to touch astronomy.") Hubble's letter announcing his discovery to Shapley was written in February 1924.

Here is the undocumented part of the story. Earlier, around 1920, when Shapley was still working at Mt. Wilson, Humason came to him one day with a photographic plate of the Andromeda nebula, which he had marked in a few places, and told Shapley that he thought the marked spots might, in fact, be

Cepheid variables. Shapley is said to have rubbed the marks away with a comment that they could not be stars since the nebula is solely a spiral of hot gases. Not surprisingly, upon reading Hubble's letter in 1924, Shapley is reported to have remarked: "Here is the letter that has destroyed my universe."

Owen Gingerich, a colleague of Shapley's at Harvard who knew the story, asked him about it in the 1970s, and Shapley replied—although by now his memory was less than reliable—that the story was possibly true. If so, then Shapley literally rubbed out one the great discoveries in astronomy of the last century. Had he not, that wonderful telescope orbiting our Earth as I write this might have been called the "Shapley" (not "Hubble") space telescope.

12.4. Einstein Defends His Cosmological Constant

Einstein's cosmological model, as he presented it in 1917, had no immediate relevance to the work of observational astronomers. Yet a Dutch astronomer, Willem de Sitter, interpreted Einstein's cosmic equation differently, producing a model that was still fundamentally static but that entailed an apparent recession of all the stars. Due to the disruption in scientific communication during the First World War, de Sitter was not aware of Slipher's work on the redshift of the nebulae. But de Sitter did correspond with Einstein on the model. Einstein realized that an expansion of the stars, whether real or apparent, contradicted the stability assumption of his paper. After several exchanges between them, Einstein remained committed to a stable universe. In the chronology of the letters to de Sitter, Einstein remarked that de Sitter's model "corresponds to no physical possibility"; that its geometrical structure "does not make sense"; and that the possible consequence of a beginning to a nonstable universe "irritates me." Note the progressively narrow-mindedness of these responses.

A second argument for a nonstable universe came forth from the Russian mathematician, Aleksandr Friedmann, in two papers (1922 and 1924). The papers were on purely mathematical aspects of Einstein's equations of 1917, with no physical applications; nevertheless, Friedmann's result showed that a nonstatic solution, which ignored the cosmological constant, was logically possible. Einstein again balked at the implication. Only a stable model was compatible with his conception of the universe.

A third assault on his model came in 1927. Unaware of Friedmann's work, the Belgian physicist Georges Lemaître showed that an expanding universe was implied if one eliminated the cosmological constant, and he offered the red shifts of the nebulae as evidence. Einstein remained adamantly opposed to a nonstatic universe. He and Lemaître met in Brussels in October 1927 at a scientific conference, where he told Lemaître of Friedmann's work, but of course Einstein had not changed his mind about either of them. Lemaître later recalled one of Einstein's comments (which echoed the exchange with de Sitter): "Your calculations are correct, but your physical insight is abominable."

On his trip to California in the winter of 1930–31, Einstein was accompanied by his second wife, Elsa, his secretary, and an assistant. On December 30, 1930, Einstein and his entourage sailed into San Diego harbor to much fanfare, for by now he was a scientific celebrity. They were taken to Pasadena the next day where, at the California Institute of Technology (Caltech) and the Mt. Wilson Observatory, he would spend the next two months. He had been invited by Robert A. Millikan, president of Caltech, whose experimental work on the photoelectric effect had confirmed Einstein's 1905 prediction of the particle nature of light; for this, plus his measurement of the electron's charge, Millikan received the Nobel Prize in 1923. Ironically, Millikan's experiment on the photoelectric effect was initially performed to disprove Einstein's model, and even when the results confirmed it, Millikan remained skeptical of the theory, because of the efficacy of the wave theory of light; in addition, he had doubts about the 1919 solar eclipse experiment as confirming Einstein prediction of the bending of light around the sun, saying that the apparent deflection of light may be due to refraction by gases around the sun. Nevertheless, with these annual winter visits, Millikan was trying to attract Einstein to a permanent position at Caltech. Einstein came a second time (1931–32), and during a third visit (1932–33) the Nazis came to power and he never returned to Germany. Ultimately, despite Millikan's pursuit of Einstein, Caltech lost to the newly created Institute for Advanced Study in Princeton, New Jersey, where Einstein remained for the rest of this life (1933–1955).

During Einstein's first visit to Caltech in 1931, he met other physicists and particularly astronomers associated with the observatory. There was Walter S. Adams, who worked on the gravitational red shift of the star Sirius B. General relativity predicted that large gravitational fields from massive stars should cause escaping light to be lengthened, and hence shifted to the red end of the spectrum (a deduction that, importantly, is independent of the "other" redshift due to motion). Charles E. St. John was also looking for this gravitational redshift from the sun. William W. Campbell had measured the bending of light during the 1922 solar eclipse, confirming further (beyond the famous 1919 experiment) Einstein's prediction (despite, like Millikan, being initially skeptical of the theory).

Most consequential was his meeting with Hubble and his assistant Humason; 1931 was a propitious time for them because they were preparing their second paper with further data confirming the linear relationship between redshifts and distances of the nebulae. The key question was: Are the redshifts Doppler shifts? Hubble toyed with this idea at the time. Ultimately he abandoned it. He was a stanch empiricist, and believed that the redshift–distance relationship (which was not called Hubble's law until the 1950s) was no more than an empirical correlation, providing a means to measure distances beyond the Cepheid limit; to Hubble anything else was mere speculation, not hard science. This apparently was his assessment even at his death in 1953 (see section 5.4).

What of Einstein in 1931? What did he take from his meetings with Hubble? I have not been able to find any documents from him, nor from the other scientists, on his view at the time. Sadly his diary addresses mundane matters. But Einstein was then a celebrity, and thankfully the *New York Times* had a reporter snooping around Caltech writing almost daily dispatches, from which I've been able to glean these remarks. In early January Einstein is quoted as saying: "New observations by Hubble and Humason . . . concerning the redshift of light in distant nebulae make the presumptions near [i.e., make it appear likely] that the general structure of the universe is not static." Here is the first intimation from Einstein of his budging from his steadfastness to a stable/static model. Then in early February the *Times* reported that he announced at a lecture that he dropped the idea of a closed (and, hence, stable) universe. A week later he confessed, "The redshift of distant nebulae has smashed my old [theoretical] construction like a hammer blow," and at the lecture he said this in an animated manner while "swinging down his hand to illustrate." No doubt that swing of the hand had a cathartic effect, liberating Einstein from his stubborn allegiance to the stable model and its cosmological constant. He was much later quoted as saying that the postulation of the cosmological constant was "the greatest blunder of my life."

I therefore suspect that at some time during this crucial visit to Caltech the thought must have occurred to Einstein that if he had not been so quick to introduce the cosmological constant, and instead explored the consequences of the equations of general relativity without it, he may have predicted the expanding universe, independently of and *before* de Sitter, Friedmann, and Lemaître, which surely would have been a crowning achievement in his already illustrious career.

In light of the "book-ends" chapters of this book (1 and 12), Einstein's tenacity and stubbornness were surely a double-edged sword—at once a source of success and failure in his scientific life.

As a coda to this chapter, I should mention that in recent years a hypothesis about an inflationary element in the expanding universe has been gaining favor. Empirical evidence is mounting that the rate of expansion (over about the last 5 billion years) has been increasing, and hence there is the requisite postulation of an additional energy (called dark energy) within the vacuum of space to account for it. This dark energy acts like a repulsive force (counter to gravity) and thus, not surprisingly, reference is often made to Einstein's cosmological constant as a forerunner of this. Perhaps Einstein's "blunder" was not so far off?

Galaxies and the Naked Eye

In the night sky, how far can we see with the naked eye? Almost everything we see is, in fact, within our Milky Way.

All the nebulae that are (gaseous) nebulae are in our galaxy. The nebula in the Andromeda constellation, seen in the Northern Hemisphere, is a galaxy.

So are the two clouds of Magellan, visible in the Southern Hemisphere. Despite present-day light pollution, some observers claim they can see the Spiral in Triangulum, another galaxy near Andromeda. These nearby galaxies constitute what is called our local group. Otherwise everything else we see with the naked eye at night is within our Milky Way.

Of course, a supernova in another galaxy could be visible, as was the one appearing in 1987 in the large cloud of Magellan. Two of the famous novae in history, so-called Tycho's nova of 1572 and Kepler's of 1604, however, were both within our galaxy.

Notes and References

Einstein's 1930–31 visit to Caltech is discussed (with remarks on Hubble's view of his law) in David Topper and Dwight Vincent, "Posing Einstein's Question, Questioning Einstein's Pose," *The Physics Teacher,* 38, No. 5 (May, 2000), pp. 278–288. On Einstein's exchange with de Sitter, see Carla Kahn and Franz Kahn, "Letters from Einstein to de Sitter on the Nature of the Universe," *Nature,* 257 (October 9, 1975), pp. 451–454. The Lemaître quotation from Einstein is in Robert W. Smith, "Edwin P. Hubble and the Transformation of Cosmology," *Physics Today* (April, 1990), pp. 52–58 (p. 57). An important article on 20th century cosmology is Helge Kragh and Robert W. Smith, "Who Discovered the Expanding Universe?," *History of Science* 41, No. 2 (June, 2003), pp. 141–162.

On the story of Shapley's blunder and the Gingerich anecdote, see Robert W. Smith, *The Expanding Universe: Astronomy's "Great Debate" 1900–1931* (Cambridge: Cambridge University Press, 1982), p.144, note 122. Also, Shapley's remark upon hearing of Hubble's discovery is quoted in Noriss S. Hetherington, "Hubble's Cosmology," *American Scientist* 78 (March–April, 1990), pp. 142–151 (on p. 144). Shapley's autobiography is, *Through Rugged Ways to the Stars: The Reminiscences of an Astronomer* (New York: Charles Scribner's Sons, 1969).

Einstein's famous statement about the cosmological constant being the biggest blunder of his life is reported by the physicist and cosmologist George Gamow; he mentions it in an article "The Evolutionary Universe," in *Scientific American* (September, 1956) and repeats it in *My World Line: An Informal Autobiography* (New York: Viking Press, 1970), p. 44.

Postlude

One of the themes of this book has been the disjunction between, on the one hand, the image of science that appears in science courses and textbooks, and, on the other hand, that uncovered by a study of the actual history. As seen, there is a sharp distinction between the complex and often-muddled process of wrestling with nature in the search for order (which the scientist pursues and which the historian subsequently probes and tries to reconstruct) and the selective image that emerges after nature, so to speak, reveals its secret. The uncovered order usually appears as an abstract, succinct, and elegant summary of a facet of nature, and is presented thusly in textbooks.

Regretfully, the alternate (historical) depiction of science sometimes entails aspects of the behavior of scientists that is less than exemplary, as was seen in many of the stories told here. Ideals and idols may become suspect and tarnished by history, for there is an iconoclastic side to the history of science—indeed, this is true of a good deal of historical work. Because of this there has arisen the accusation that historians are out to debunk science, that they are aligned with the so-called postmodernists, some of whom assert that all knowledge—including, and especially, Western scientific thought, with its claim to objectivity—is actually a subjective enterprise grounded in and reflective of the embedded culture (sometimes referred to as "socially constructed"), and is thus on equal footing or of equal truth value with other "ways of knowing," such as astrology or psychic experiences. This viewpoint has given "contextualism" a bad name. On the other hand, and in reaction to postmodernism, there has arisen a coterie of very vocal science zealots, aggressively defending the hegemony of science seemingly over all matters of knowledge. Whereas the postmoderns tend to be naïve about science, the zealots, I find, are naïve about the *history* of science. Perhaps they are partially blinded by the simple beauty of science as seen only through the textbook.

I admit that this book may raise serious doubts about the objectivity of some specific episodes told here. As seen, I confess to being a skeptic, yet a skeptic grounded in the study of history (recall the short essays on skepticism in Chapter 5). At the same time, I am convinced there is truth-content in the study of history as in science. The historical records provide the empirical constraints within which I work, these being analogous to epistemological constraints entailed by the natural world for the scientist. The following quotation by Richard Jenkyns, professor of the classical tradition at Oxford, splendidly concurs with my viewpoint on historical methodology:

Somehow we need to keep in balance our sense that history-writing is a creative and imaginative process with our belief that the good historian is in search of truth and understanding. The historian's personal character and beliefs are instruments that can be used well or ill. [Edward] Gibbon's *Decline and Fall [of the Roman Empire]* is pervaded by his skepticism: it is one of the sources of his greatness, but there are occasions when it led him to distortion, and at those moments he was the lesser historian. We are not purely at the mercy of our prejudices: every scholar has known the times when the evidence has commanded him to abandon a position dearly held. At such a moment the bad historian cheats; the good one reconsiders, even at the cost of demolishing what he has built. But if there were nothing to the inquiry but the free flow of subjectively, such a dilemma could not arise.

I have faced that dilemma many times in my work. In my research on Newton and the colors of the spectrum (Chapter 9), I initially set out to show that the musical metaphor was the source of the seven colors, but the primary documents forced me to see that the aesthetic argument came first, with the musical analogy following. The same occurred in my research on Galileo and sunspots. I wanted to vindicate Galileo from some nefarious accusations made by Arthur Koestler in his popular book, *The Sleepwalkers*. In the course of finding an error in Koestler and hence vindicating Galileo, I uncovered Galileo's less than honest exposition of the role of the precession of the equinoxes (see section 4.4). As a historian, I try to reconstruct what really happened, akin to a scientist looking for what the world is really like. Otherwise, why bother?

In the long run, I believe, science is the best and most powerful means we have of fathoming the external world. In the short run, however, the fundamentally all-too-human and less-than-rational characteristics of scientists are more manifestly obvious, revealing, to use a phrase, their quirky side.

Notes and References

I discuss Koestler's critique of Galileo in "Galileo, Sunspots, and the Motions of the Earth: Redux," *Isis,* 90 (1999), pp. 757–767. The quotation by Jenkyns is from his essay review, "The Labyrinth of Arthur Evans," in *The New York Review of Books* (November 1, 2001), pp. 65–67.

Index

Entries are filed word-by-word. Locators followed by an asterisk (*) indicate figures; those followed by *t* indicate tables. **Boldface** locators indicate the more principal treatments of those topics with extensive citations.